AF606321

HEPARIN (AND RELATED POLYSACCHARIDES)

Polymer Monographs

Edited by M. B. Huglin, University of Salford

A series of short monographs, each dealing with one specific polymer. They are written by experts from leading research laboratories, and cover equally basic scientific information on a polymer and information pertinent to its practical utilization.

International Standard Serial Number (ISSN): 0275–5777.

Volume 1
POLY(1-BUTENE)—Its preparation and properties
I. D. RUBIN, *Texaco Inc.*

Volume 2
POLYPROPYLENE
H. P. FRANK, *Osterreichische Stickstoffwerke*

Volume 3
POLY(VINYL CHLORIDE)
J. V. KOLESKE and L. H. WARTMAN, *Union Carbide Corporation*

Volume 4
POLY(VINYL ALCOHOL)—Basic properties and uses
J. G. PRITCHARD, *West Ham College of Technology*

Volume 5
POLYVINYLIDENE CHLORIDE
R. A. WESSLING, *Dow Chemical Co.*

Volume 6
POLY(N-VINYLCARBAZOLE)
J. M. PEARSON, *Eastman Kodak Co.* and M. STOLKA, *Xerox Corporation*

Volume 7
HEPARIN (AND RELATED POLYSACCHARIDES)—Structural and Functional Properties
WAYNE D. COMPER, *Monash University*

Volume 8
POLY(TETRAHYDROFURAN)
P. DREYFUSS, *University of Akron*

The publisher will accept continuation orders which may be cancelled at any time and which provide for automatic billing and shipping of each title in the series upon publication. Please write for details.

HEPARIN (AND RELATED POLYSACCHARIDES)

Structural and Functional Properties

WAYNE D. COMPER
Monash University

GORDON AND BREACH SCIENCE PUBLISHERS
New York London Paris

Gordon and Breach, Science Publishers, Inc.
One Park Avenue
New York, NY 10016

Gordon and Breach Science Publishers Ltd.
42 William IV Street
London, WC2N 4DE

Gordon & Breach
7–9 rue Emile Dubois
F-75014 Paris

Library of Congress Cataloging in Publication Data

Comper, Wayne D.
Heparin (and related polysaccharides)

(Polymer monographs; v. 7)
Includes index and references.
1. Heparin. 2. Polysaccharides. 3. Structure-activity relationship (Pharmacology) I. Title. II. Series.
QP702.H4C65 591.1'13 81-4257
ISBN 0-677-05040-2

To My Family

Contents

Series Editor's Preface

"*If a book is worth reading, it is worth buying.*" (John Ruskin)

Two trends, which are not especially favourable to a series such as this have become noticeable in recent years, *vis*—first, the reluctance of the polymer industry to invest in the development of entirely new homopolymers and, secondly, the increasing costs of scientific book production. Enquiry and feedback revealed, perhaps surprisingly, but nonetheless emphatically, a genuine need for authoritative up-to-date treatments on several existing polymers. Consequently, each volume is devoted to one specified polymer. The interests of an industry or an institution usually dictate that research and/or development be conducted over an extended period of time on the particular polymer of relevance. It is hoped that individuals will be able to select their specific volume from Polymer Monographs and thus be freed from the inconvenient and unfair obligation to pick sections out of the commoner, large treatises devoted to classes of polymer.

With regard to the topics themselves, proven useful application is considered a *sine qua non*, and hence macromolecules of purely academic interest are excluded from this series. An innovation is the inclusion of biopolymers and their synthetic prototypes. Since polymer science has now become truly interdisciplinary in scope, the monographs are addressed to a broad spectrum of potential readers.

These volumes are short, the aim being to present the maximum of current information in the minimum of space. Fortunately the authors, who are all pre-eminent in their respective fields, have not only complied with this difficult and stringent condition, but have succeeded in doing so without sacrificing readability. The publishers have endeavoured to make these books available at a price which is reasonable by present standards. They and I trust these volumes will prove useful and welcome comments and suggestions for future topics.

MALCOLM B. HUGLIN

Acknowledgements

My scientific life owes much to those with whom I have had the privilege to work. It is with great pleasure that I thank Assoc. Prof. Barry Preston for his support, encouragement and friendship, not only in my formulative years as a PhD. student, but as a present-day collaborator at Monash. I am also most indebted to Prof. Torvard Laurent (Uppsala University) and Prof. Arthur Veis (Northwestern University) for giving me the opportunity to work in their laboratories for postdoctoral studies. I can only hope that my attempt to write the book embodies to some extent the high standards and quality of work that these people attain.

I am further indebted to Prof. Laurent for giving me the opportunity to write the book. His support and expert critical comments throughout the final stages of the writing have been invaluable. For expert critical comments on various chapters of the book, I am indebted to Dr J. J. Hopwood, Prof. U. Lindahl, Assoc. Prof. B. N. Preston, Dr H. C. Robinson and Dr R. D. Rosenberg. Certainly, any errors in style and interpretation are of my own doing.

I am also extremely grateful to Profs. I. Bjork, U. Lindahl and R. D. Rosenberg for sending me preprints of their submitted and unpublished work. This information was invaluable as much of their work highlights many sections of the book.

To the members of the Biochemistry Department, Monash University, I am grateful for their support throughout the course of writing this book. In particular, it is a great pleasure to have the opportunity to thank Diane Mathers and John Ford for their invaluable help and their friendship. I thank Anne Heffernan, who did a mighty task of typing the book and managed to live and type through my excruciating moments when finding my own errors. I also thank Linda Masters for helping with typing corrections. To my mother, I can only say that the perseverance, persistence and other talents required to write the book are miniscule compared to her effort in life. I am indebted to Rowly Dolman for all his help. I am most grateful to my parents-in-law for their tremendous help in our accommodation during periods of writing. And to my wife, who has suffered an author's emotional traumas, isolationism and other typical author's human responses when performing this exercise, and who has had two magnificent sons, Albert and Lee, one born at the commencement of writing this book, and the other on its completion (perhaps two of an author's typical human responses), I can only say her effort has been supreme.

Preface

Why write a book on heparin? It has intrigued me that, whereas our knowledge of the chemistry and structure of connective tissue polysaccharides (including heparin) is considerable, our knowledge of the biological role of these polysaccharides is far from clear. It is primarily for this reason that writing a book on heparin may satisfy my curiosity concerning this problem. Given the existence of a missing link between structure and function of heparin is always good material for those philosophically bent, it is clear that other problem-solving strategies should be adopted. A comprehensive treatment of the properties and interactions of heparin in a unified manner may help define the way.

The high quality work in areas of heparin anticoagulant activity, structural analysis, metabolism and interaction with cell surfaces has involved an explosive output of work in the last five years. These results and new concepts are still settling. In this arena of transience a book on heparin becomes more a newspaper article. The anticipated headlines will outdate the book quickly. However, in describing the events of the last decade I have tried to place, in parallel, recent results on heparin with its chequered past and sometimes hackneyed concepts. This may place the many diverse aspects of heparin in proper context, clear up apparent contradictions, reduce confusion and improve the chance of giving new insights into heparin.

I have placed major emphasis on the structural and functional aspects of endogenous heparin and related polysaccharides. Many important aspects of their clinical and pharmacological usage have not been included, except where it is considered appropriate that they focus on the question of the properties of endogenous material. I have frequently drawn upon many excellent review articles and the reader of both will find similarities. On the other hand, I have attempted to draw together and document much of the fragmentary and isolated information associated with the physicochemical properties of these polysaccharides.

WAYNE D. COMPER

Section A

General Introduction

CHAPTER 1

Introduction

1.1 BRIEF HISTORY

The property of heparin, which has attracted most attention and resulted in its universal clinical use, is its anticoagulant activity. As a result, heparin is usually defined as an anticoagulant (as in most general reference works).

Attention was drawn to this property at the turn of the century where a number of workers (Pavlov, 1887; Schmidt, 1892; Morawitz, 1905; Doyon, 1912 cited in Best (1959) and Jaques (1978a)) had isolated preparations with anticoagulant properties. The significance of these findings was enhanced by the work of Jay MacLean [1891–1957] who, while working in Howell's laboratory as a medical student, set out to purify thromboplastic active materials previously reported by Howell (1912) to be due to cephalin-protein (MacLean, 1916). It was during this investigation that MacLean became interested in the deterioration of these preparations from various tissues as evidenced by their loss of thromboplastic activity. It was found that several preparations had not only lost their thromboplastic action, but actually achieved anticoagulant status. In 1918, Howell and Holt extended this study and called the material from liver, heparin derived from the Greek word *hepar* meaning liver. Isolation consisted of ethereal extraction and purification by repeated reprecipitations of the ether solution with acetone and then absolute ethanol. The antagonism between cephalin (thromboplastic active) and heparin on the clotting system was described in the Howell and Holt paper. However, there is some controversy as to whether preparations by MacLean, Howell and Holt did actually represent heparin (Jaques, 1978a) although MacLean is generally accredited with the discovery of heparin. Serendipity did not apparently play a role in MacLean's 'discovery' as he states later in correspondence (Best, 1959) "Naturally I regard the statements in the literature that I discovered this (*sic* anticoagulant) 'accidentally' as not correct. It was discovered 'incidentally' in the course of the problem but not 'accidentally.'"

In 1922 Howell described under the same name 'heparin' an anticoagulant material prepared from dog liver by aqueous extraction and acetone precipitation, and in the period between 1922 and 1928 demonstrated its carbohydrate nature and the fact that it contained sulphur.

Heparin was first isolated from beef lung in sufficient purity for clinical use and for chemical study by Scott and Charles (1933). They demonstrated that it was an acidic carbohydrate capable of forming salts with metals and that it was reasonably free of other substances by crystallizing it as a barium salt (Charles and Scott, 1936). The identification of uronic acid (Schmitz and Fischer, 1933), sulphate (Jorpes, 1935) and glucosamine (Jorpes and Bergström, 1936) as the main components demonstrated that heparin was a highly sulphated anionic polysaccharide.

The realization that heparin preparations exhibited complex chemical heterogeneity was made as early as 1948, when Jorpes and Gardell isolated a by-product from a commercial preparation of heparin, which showed a positive optical rotation, but differed from heparin by the presence of acetyl groups and had low anticoagulant activity. They concluded that the substance, which they called heparin monosulphuric acid, contained one equivalent disaccharide of hexosamine, uronic acid, sulphate and acetyl moieties and thereby constituted material different from heparin. Similar material was isolated by Meyer *et al.* (Meyer *et al.*, 1956; Linker *et al.*, 1958) from the liver of a patient with amyloidosis and from other tissues and was called heparitin monosulphate. These substances were later to be more commonly known as heparan sulphates (Table 1.1).

1.2 HEPARIN AND HEPARAN SULPHATE

Heparin is the highest negatively charged molecule in tissues. It is widely distributed in mammalian tissues and fluids, the largest

TABLE 1.1
Nomenclature of the Glycosaminoglycans and Proteoglycans

Contemporary version	Other names	Abbreviation
glycosaminoglycan	glycosaminoglycuronan	GAG
	acid mucopolysaccharide	
	acidic mucopolysaccharide	
	sulphated mucopolysaccharide	
	amino polysaccharide	
	mucopolysaccharide	
chondroitin 4-sulphate	chondroitin sulphate A	C4-S
	chondroitin A	
chondroitin 6-sulphate	chondroitin sulphate C	C6-S
	chondroitin C	
dermatan sulphate	chondroitin sulphate B	DS
	chondroitin B	
	B-heparin	
hyaluronate	hyaluronic acid	HA
keratan sulphate (1)	corneal keratan sulphate	KS (1)
keratan sulphate (11)	skeletal keratan sulphate	KS (11)
	keratosulphate	
	keratan	
heparin-like polysaccharides		
heparin	α-heparin or porcine heparin	Hep
	ω-heparin or whale heparin	
heparan sulphate	heparitin sulphate	HS
	heparitin sulphate (A, B, C, D)	
	heparin monosulphate	
	heparin sulphate	
	N-acetylheparin sulphate	
proteoglycan	mucopolysaccharide-protein complex	PG
	mucoprotein	
	polysaccharide-protein	
	protein polysaccharide	
multi-chain heparin	macromolecular heparin	
	high-molecular weight heparin	
	heparin proteoglycan	

amounts being in lung, spleen, liver and muscle. It is the only major representative of the glycosaminoglycan group in mammals which exists normally in an intracellular environment, apparently contributing to the frame-work of the granular cytoplasmic inclusions of mast cells. The mast cells are located in connective tissues alongside the capillaries and in the wall of blood vessels and in other loose connective tissues. On the other hand, the other vertebrate glycosaminoglycans, including heparan sulphate, have a largely extracellular distribution in the amorphous extracellular matrix of connective tissues or, as with heparan sulphate, are attached to the cell surface. The non heparin-like glycosaminoglycans generally occur in much higher amounts in tissues as compared to the 'trace' nature of heparin and heparan sulphate.

The classical differences between heparin and heparan sulphate were registered on the basis of sulphate content and anticoagulant activity (heparin having relatively higher sulphate and anticoagulant activity). In broad terms, low sulphated, D-glucuronic acid-rich polysaccharides are classified as heparan sulphate, whereas high sulphated, L-iduronic acid-rich species are designated heparin. In more recent times, Lindahl (1976) has suggested that there are no sufficient criteria to clearly indicate differences between these molecules on a structural basis particularly for intermediate-type polymers and that both molecules represent the extremes for a spectrum of molecules. In view of this it seems more appropriate to consider both polymers as members of the same family, i.e. 'heparin-like polysaccharides' or heparin and related polysaccharides. However, the reader will find throughout the book references to heparin and heparan sulphate. This is not by choice (as it is confusing), but is used only to relay some continuity with published work. The clearest distinction between these two molecules is apparent in the amino acid composition of the protein cores of the proteoglycan or multi-chain forms of these polysaccharides (see Section 4.3). This distinction is hardly practical at the present time, as normal preparations are merely polysaccharide chain preparations in which the multi-chain forms have been degraded. It is not known whether the different protein cores of heparin and heparan sulphate do actually distinguish or determine the type of polysaccharide chain synthesized.

As the study of heparin is intimately related to that of the other glycosaminoglycans, it is pertinent to consider briefly some particular and general aspects of chemical structure in this group of compounds.

1.3 OTHER GLYCOSAMINOGLYCANS

The structures of the different types of vertebrate glycosaminoglycans are similar in that they may generally be described as unbranched polymers built of disaccharide repeat units consisting of hexosamine and hexuronic acid (with the exception of keratan sulphate, which has glucosamine and galactose). Generally, the glycosaminoglycans do not occur as free polysaccharide chains *in vivo*, but as proteoglycans in which several chains are covalently linked at the reducing terminal to a polypeptide core. Hyaluronate may be an exception, as it is not clear whether it exists in a proteoglycan form.

Earlier views of glycosaminoglycan structures generally presented an idealized or 'parent' type of structure with a continuous repeat of a single disaccharide unit. It is becoming increasingly evident, however, that this representation is too simple and that actual molecules rarely contain only a single type of disaccharide unit. The parent type disaccharide units of the five non-heparin-like glycosaminoglycans are shown in Figure 1.1. The structures for the heparin-like glycosaminoglycans will be discussed in Chapter 2.

Hyaluronate seems to represent the simplest type of glycosaminoglycan structure in which only one known disaccharide repeat unit is found.

The chrondroitin sulphates generally have rather simple structures, although the individual chondroitin sulphate chains may contain sulphate groups in both 4– and 6–positions but may be exclusively 4–sulphated or 6–sulphated.

The iduronate-containing polymers, namely dermatan sulphate and the heparin-like polysaccharides, may contain a large number of different disaccharide units, arranged either in block structures or in less well-ordered, complex sequences.

The glycosaminoglycans are then specifically defined by their fine topochemistry i.e. their content and distribution of charged anionic groups. The other factors of the polysaccharide chain, which are of importance in determining configurational properties, are the hydroxyl and N-acetyl groups, and the position and configuration of the glycosidic linkages.

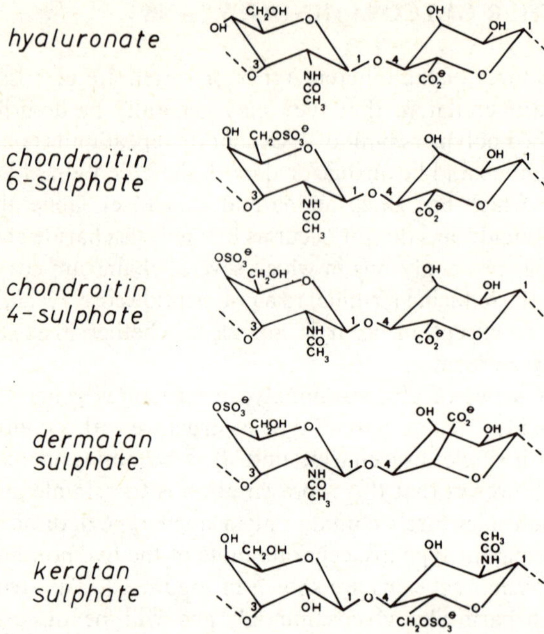

FIGURE 1.1 Dominant disaccharide unit of the five non heparin-like connective tissue glycosaminoglycans.

1.4 NOMENCLATURE

The nomenclature for the group of glycosaminoglycan substances has been quite variable and certainly inconsistent. The basic trivial names of the glycosaminoglycans were derived from the tissue from which the material was originally prepared. The trivial names have undergone a great deal of variation, generally according to the particular fractionation and chemical properties of preparations made in various laboratories. A nomenclature based on their component monosaccharides has proved difficult, even presently, due to the intrinsic chemical heterogeneity exhibited by these compounds.

For the sake of clarity Table 1.1 outlines the nomenclature to be used to describe this group of compounds in this text together with names (old and new) that have been used elsewhere. No attempt has been made to introduce a new nomenclature here but use is made of names which have popular present-day usage and which are primarily based upon the nomenclature suggested by Jeanloz (1960). More recent attempts at nomenclature classification have been made by Jaques (1977, 1978b). These are based on describing the glycosaminoglycan by its trivial root name (e.g. chondroitin, heparitin) followed by an alphabetical letter to designate characteristic chemical differences within a particular glycosaminoglycan group.

The term 'heparinoids' has been used chiefly for the products of sulphated or resulphation of polysaccharides (usually after some depolymerization). The term has had this association, because the compounds were originally prepared for testing as substitute anticoagulants.

1.5 GENERAL STRUCTURAL AND FUNCTIONAL ASPECTS OF CONNECTIVE TISSUE POLYSACCHARIDES

In view of the extensive development of physicochemical, fractionation and tissue and cell culture techniques that has occurred over the last 15 years or so, a large and disproportionate amount of information has been gained on the chemical characterization and structure of glycosaminoglycans and proteoglycans. While it is true that this approach maintains popularity in this field, attention directed to the functional properties of these molecules has been scant. The assignment of functions to these molecules, particularly in relation to their structure, has proved a neglected, if not an easy, task. In effect, we have 'structures in search of functions'. Therefore, if we are to understand why these molecules exist and the functions they perform, then the concepts of structural–functional relationships and of specific biological activity loom large. Various approaches to the functional aspects of glycosaminoglycans and proteoglycans have been taken so far.

The fact that glycosaminoglycans occur in different connective tissues in defined typical patterns which exhibit distinct regimes of change with maturation and aging, is well established. These topographical aspects, together with the observed variation of the type and distance of charged anionic groups along the glycosaminoglycan chain, led Meyer (1960) to suggest that they 'act as templates or organizers in the development of the architecture of the organized fibrous elements of connective tissues'. As yet there is not enough information to suggest that these molecules demonstrate 'extrachromosal codon' capabilities to perform this organizing task, although there are several lines of evidence to support such a concept (Section C).

More recently, two different structure–function approaches have emerged. One view has been taken to explain the functional properties of the glycosaminoglycans and proteoglycans in terms of their overall macromolecular and polyelectrolyte properties e.g. size, shape and charge density; these properties being generated by the cooperativity of their elemental primary structures to give higher order levels of structure. The macromolecular and polyelectrolyte nature of these molecules have been used to explain the physicochemical properties they can confer on the systems and structures that contain them (Comper and Laurent, 1978). Such interpretations are pertinent to the study of a number of factors, such as:

(1) systems which contain relatively concentrated solutions or phases of these polysaccharides which generally exhibit thermodynamic non-idealities and

(2) hydrodynamic or non-equilibrium phenomena associated with polysaccharide containing systems.

The other view is one which is focused at the primary structural level of these molecules and lies in the shadows of Meyer's original hypothesis. This approach equates biological activity and interaction to the fine structural and chemical details of these macromolecules. To be sure, it is viewed that the nature of structural complexity and variability exhibited by these molecules, particularly as witnessed by molecules such as heparin and heparan sulphate, warrants such an approach (Lindahl and Höök, 1978). This functional concept is more amenable to discussing systems containing 'dilute' concentrations of glycosaminoglycans, although it is not necessarily restricted to this as such. It is probable that these two approaches are mutually complementary, that they overlap in certain respects and both are valid in structure–function interpretations.

Some further general considerations of these molecules are noteworthy. These macromolecules fulfil structural and other non-enzymatic functions which depend, in part, upon interactions with other molecules inside and outside the cell. These interactions generate higher orders of molecular organization at various levels. Their presence, whether it be an intracellular location or more commonly as an extracellular substance in the extracellular matrix of cells, presumably filters and censors, by a pseudo-catalytic mechanism, the intrinsic molecular and physical information that reaches and affects the cells. This information ultimately influences cell function in terms of viability, growth, repair and reconstruction.

1.5.1 Biological specificity

A realization of what the term specificity means is essential in understanding structure–function relationships and particularly the ramifications of what numerous investigators purport observed phenomena to the realms of specificity. A basis of analysis of biological systems is a description of the system in terms of a particular set of components giving rise to a particular set of reactions. The successive integration of these reactions and interactions yields the overall properties of the system or its 'biological function'. The term 'biological specificity' can legitimately be introduced to describe the various aspects of these particular components and reactions of the system. Although a conceptual view of these 'specific' reactions may encompass a broad stratification of the system (an intrinsic problem referred to in Chapter 8 in terms of defining 'functional' and 'structural' parts), it needs only to be stressed that biological specificity must be within the regime of the *in vivo* system. Reactions which may exhibit apparent specificity *in vitro* or by artificial manipulation of an *in vivo* system may or may not be relevant in terms of biological specificity.

Although the overall concept of 'biological specificity' may be clear, it is hardly a practical view to be taken at the moment considering our limited knowledge of structure–function relationships of glycosaminoglycans and their organization in biological systems. A more common view of specificity is limited to only particular reactions, the majority of which come under the heading of 'binding'. For example, a common mechanism which evokes the specificity concept is the 'lock and key' mechanism, which operates for antibody–antigen complexes and enzyme–substrate complexes. The complexes that are formed are highly selective with forces involved in binding (e.g. electrostatic, hydrophobic, hydrogen binding, van der Waals, etc.) being maximized over short distances. Selectivity results from the fact that their components cannot be substituted for by any other molecular species to register, in most cases, a highly 'unusual' reaction. Scott (1973) states, specificity is 'a maximum of affinity manifest in the minimum number of ways'. Although a number of forces may be involved in binding together with stereochemical aspects of the interacting species, the nature of the specificity is generally characterized by the components that participate in the reaction rather than the mode of the specific mechanism of binding. However, it is obvious that not all biological reactions are so clear cut. More subtle variations that affect macromolecular activity such as changes in conformation, association and dissociation of subunits, stabilization or protection, inhibitor and feedback interactions, may operate on a biological level. Under such conditions, our ideas of specific binding (which are stimulated through such favourable observed complexes formed *in vitro*) may have to be moderated when we consider the actual biological environment with which these reactions take place. The importance of realizing specificity at the various levels in biological systems is its relation to the properties of the system and its specific 'function'.

REFERENCES

Best, C. H., *Circulation* **19**, 79 (1959).

Charles, A. F. and Scott, D. A., *Biochem. J.* **30**, 1927 (1936).

Comper, W. D. and Laurent, T. C., *Physiol. Rev.* **58**, 255 (1978).
Howell, W. H., *Am. J. Physiol.* **31**, 1 (1912).
Howell, W. H., *Am. J. Physiol.* **63**, 434 (1922–3).
Howell, W. H. and Holt, E., *Am. J. Physiol.* **47**, 328 (1918).
Jaques, L. B., *Meth. Biochem. Anal.* **24**, 203 (1977).
Jaques, L. B., *Semin. Thromb. Haem.* **4**, 350 (1978a).
Jaques, L. B., *Artery* **4**, 114 (1978b).
Jeanloz, R. W., *Arthritis Rheum.* **3**, 323 (1960).
Jorpes, J. E., *Biochem. J.* **29**, 1817 (1935).
Jorpes, J. E. and Bergström, S., *Z. Physiol. Chem.* **244**, 253 (1936).
Jorpes, J. E. and Gardell, S., *J. Biol. Chem.* **176**, 267 (1948).
Lindahl, U., in *MTP International Review of Science. Organic Chemistry Series Two*, G. O. Aspinall (ed.), Butterworths, London and Boston, Vol. 7, Chapter 9, p 283 (1976).
Lindahl, U. and Höök, M., *Ann. Rev. Biochem.* **47**, 385 (1978).
Linker, A., Hoffman, P., Sampson, P. and Meyer, K., *Biochim. Biophys. Acta* **29**, 443 (1958).
McLean, J., *Am. J. Physiol.* **41**, 250 (1916).
Meyer, K., in *Molecular Biology. Elementary Processes of Nerve Conduction and Muscle Contraction.* D. Nachmansohn (ed.), Academic Press, London and New York, p 69 (1960).
Meyer, K., Davidson, E. A., Linker, A. and Hoffman, P., *Biochim. Biophys. Acta* **21**, 506 (1956).
Schmitz, A. and Fischer, A., *Z. Physiol. Chem.* **216**, 274 (1933).
Scott, J. E., *Biochem. Soc. Trans.* **1**, 787 (1973).
Scott, D. A. and Charles, A. F., *J. Biol. Chem.* **102**, 437 (1933).

Section B

Structure and Biosynthesis

CHAPTER 2

Monosaccharide composition

2.1 INTRODUCTION

The elucidation of the primary structure of the heparin-like polysaccharides has undergone slow progress. A number of difficulties arise in the analysis of these compounds. One of these is that their preparations, particularly those of heparin by-products which are generally designated as heparan sulphate, are not single compounds but constitute a spectrum (not necessarily continuous) of related polymers. These compounds differ, to varying degrees, in terms of the chemical adjuncts or embellishments placed on the glycose moieties of the polysaccharide chain during their synthesis. Furthermore, the preparations vary considerably with the tissue of origin and generally exhibit broad molecular weight distributions.

Indeed, it has been the employment of various fractionation techniques, utilizing differences in molecular size, polymer charge density and protein binding over the last 10 years that have clearly demonstrated the marked heterogeneity with which these molecules are endowed. The classical identification (pre-1973) of preparations as 'heparin' or 'heparan sulphate' on the basis of their chemical constituents and anticoagulant activity becomes an area of ambiguity when considering recent investigations of fractionated material which exhibit intermediate values of these two parameters.

The important questions of present concern are not only the realization of the chemical heterogeneity of these molecules but which ones are important for a particular reaction, whether it be *in vivo* or *in vitro*, to which the activity can be ascribed. This necessitates not only a fairly careful characterization of the polysaccharide fractions but also ascertainment of how the heterogeneity may vary within single polysaccharide chains rather than in a family of chains isolated from a particular source.

The sequence analysis of these structures is still in its early stages. The employment of degradation techniques, either chemical or enzymatic or both, with subsequent analysis of the various oligosaccharide products has yielded information on fractions with relatively simple structures. However, there is a lack of complete knowledge on the mechanism of action of the degrading weaponry so that more complicated structures still 'wait' to be examined sequentially.

A prime concern in this thesis is the nature of the highly variable primary structures in these polysaccharides in relation to structure–function. With this in mind, it would seem informative to summarize the basic principles of the detailed structure and chemical heterogeneity that have been established. A substantial portion of the work describing analysis techniques and assignment of monosaccharide substituents has been reviewed (Jeanloz, 1975; Lindahl, 1976; Rodén and Horowitz, 1978).

The efforts at the elucidation of these structures has been concerned with a number of factors including:

1) the type of glycosidic linkage and anomery of the carbohydrate components.

2) the nature of the uronic acid component

3) the contents of ester sulphates, N-sulphate groups and their position

4) the content of N-acetyl groups

5) the chemical sequence of these variations in each molecule and finally

6) the nature of chemical heterogeneity.

2.2 MONOSACCHARIDES

2.2.1 Amino sugar

The identity of the hexosamine content of heparin has been established for many years since Jorpes and Bergström (1936) and Wolfrom *et al.* (1943) isolated 2-amino-2-deoxy-D-glucose hydrochloride from an acidic hydrolysate of the polysaccharide with a yield of 83% (Wolfrom *et al.*, 1943). The nature of the hexosamine residue was further established when Jorpes and Gardell (1948) isolated crystalline 2-amino-2-deoxy-D-glucose hydrochloride from acidic hydrolysates of heparan sulphate. More recent studies have not disputed the assignment of D-glucosamine as being the only hexosamine-type residue constituting heparin-like polysaccharides.

2.2.2 Uronic acid

Early studies demonstrated the presence of D-glucuronic acid in heparin (Wolfrom and Rice, 1946) and heparan sulphate (Brown, 1957) together with L-iduronic acid in heparin, heparan sulphate and mactins (Cifonelli and Dorfman, 1962; for review see Brimacombe and Weber, 1964). Although the presence of D-glucuronic acid in heparin was well established, general acceptance of the presence of L-iduronic acid in heparin-like polysaccharides was delayed for some years. This was the result of using hydrolytic conditions to isolate disaccharide fragments which were too drastic to isolate stable derivatives of iduronic acid (Wolfrom *et al.*, 1964; for review see Brimacombe and Weber, 1964). Subsequently, Wolfrom *et al.* (1969a) isolated L-iduronic acid from heparin as its crystalline brucinium salt. Perlin *et al.* (1970) have established by nmr techniques that L-iduronic acid is a major fraction of the total uronic acid in heparin.

Attempts to determine the uronic acid composition of heparin have been largely hampered by the resistance of these polysaccharides towards acid hydrolysis (Lindahl, 1976). This is due to the presence of N-sulphate groups in heparin which, when removed in the early stages of acid hydrolysis leads to protonation of amino groups and thereby markedly increasing the resistance of the glucosaminidic bonds to hydrolysis. Taylor and Conrad, 1972; and Taylor *et al.*, 1973 have determined the molar ratio L-iduronic and D-glucuronic acid contents of a variety of heparin and heparan sulphates by a nondestructive depolymerization technique. They measured by radiochromatography the amounts of ^{3}H recovered in glucose, idose and idosan when the polymers are carboxyl-reduced with carbodiimide and sodium [^{3}H]-

borohydride, which converts the acid-resistant hexuronidic bonds into more acid labile bonds. The reduced polymer, which is labelled with tritium atoms on the reduced carboxyl group is then stoichiometrically depolymerized by acidic hydrolysis followed by deaminative cleavage of the remaining glucosaminides with nitrous acid. The percentage of the total uronic acid represented by L-iduronic acid varied from 50 to 90% in heparins and from 30 to 55% in heparan sulphates (Table 2.1).

For the liberation of free uronic acid in high yield, Höök *et al.* (1974) have developed a method based on a combination of two procedures:

1) treatment with nitrous acid to yield uronosylanhydromannose disaccharides and some higher oligosaccharides and

2) acid hydrolysis with trifluoroacetic acid. The combined treatment of 1) and 2) yields free uronic acid to an extent corresponding to 75–80% of the total uronic acid.

One unusual feature associated with the uronic acid composition is that both standard colourimetric methods (Dische, 1947; Bitter and Muir, 1962) and the carbodiimide technique (Taylor *et al.*, 1973) yield molar ratios of uronic acid to hexosamine considerably in excess of 1.0 (Table 2.1) in spite of a large accumulation of data which has led to the belief that the basic heparin structure is best represented by the repeating disaccharide, hexuronosyl-glucosamine (for review Jeanloz, 1970, 1975). Subsequently, Shively and Conrad (1976) have demonstrated the formation of a previously unidentified 2,5-anhydrouronic acid from approximately 25% of the L-iduronic acid 2-sulphate residues during heparin acid hydrolysis (Figure 2.1). Its formation does not occur with prior reduction of the C-6 carboxyl. The amount of anhydrouronic acid formed is determined by the relative rates of the S–O and C–O bond scission prior to the hydrolysis of the glycosidic bond of the sulphated uronic acid.

When the sulphate is hydrolytically removed before cleavage of the glycosidic bond or when the glycosidic bond is hydrolyzed before loss of the sulphate the elimination reaction no longer occurs. The presence of the ester sulphate moiety at C-2 of the iduronate residue seems to promote the formation of this acid. Shively and Conrad suggest that this offers a possible explanation

TABLE 2.1 Chemical Composition of Various Heparin-like Polysaccharide Preparations with Other Physical, Structural, and Biological Parameters

Fraction	Source	Uronic Acid[a] (%)	Hexosamine[b] (%)	Mol UA[c] / Mol of GlcN: GlcN	Mol / Mol of GlcN: Gal	Mol / Mol of GlcN: Xyl	$[\alpha]_D^{24}$ [d] (deg)	Mol Wt ($\times 10^{-3}$)	DP[e]	% ot Total Uronic Acid: GlcUA Linkage[f]	% ot Total Uronic Acid: GlcUA Backbone	% ot Total Uronic Acid: IdUA Backbone	Sulphate (mol mol of GlcN): N–	Sulphate (mol mol of GlcN): O–	Anticoag. Act. (IU/mg)
Heparins															
BLH I	Beef lung	39	23		0.007	0.003	43	11	16	6	19	75	0.98	1.48	180
BLH II	Beef lung	38	21	1.4	0.005	0.001	38	11	16	6	10	84	0.96	1.14	109
BLH III	Beef lung	40	22	1.5	0.030	0.021	44	5.2	10	10	17	73	0.86	1.12	
HMH I	Hog mucosa			1.3	0.031	0.020	41	10.5	16	6	21	73	0.89	1.41	177
HMH II	Hog mucosa	36	21	1.4	0.080	0.042	41	9	17	6	36	58	0.81	1.04	114
BMH I	Beef mucosa	49	24	1.5	0.050	0.035		11	16	6	28	66	0.86	1.56	152
WH I	Whale	43	23		0.011	0.004	67	11	18	5	35	60	0.74	1.06	176
M I	Clam	39	26	1.4	0.080	0.020	47	17	31	3	50	47	0.83	0.86	132
Heparan sulphates															
HSH I	Hog mucosa by-product	38	21	1.4	0.042	0.022	61	6.2	12	9	39	52	0.63	0.64	16
HSH II	Hog mucosa by-product	45	30	1.2	0.011	0.006		16	33	3	64	33	0.34	0.38	7
HSH III	Hog mucosa by-product	34	26	1.3	0.010	0.005	67	14	28	3	70	27	0.48	0.42	<10
HSB I	Beef lung by-product	38	21	1.3			62	15.5	31	3	56	41	0.49	0.53	
HSB II	Beef lung by-product	38	25	1.3	0.042	0.024	54	17	32	3	57	40	0.56	0.82	13
HSB III	Beef lung by-product	44	25	1.2	0.028	0.018	73	16	32	3	60	37	0.51	0.48	40
HSU I	Human umbilical cord	45	29	1.2	0.042	0.020	65	27	58	2	63	35	0.32	0.22	

(Adapted from Taylor *et al.* (1973) with permission of publishers)

[a] Determined as described by Dische (1947)

[b] Determined as described by Boas (1953)

[c] Molar ratio of total hexuronic acid to hexosamine determined by radiochromatographic analysis (Conrad *et al.*, 1973)

[d] All rotations, measured at a concentration of 1% in water, are positive.

[e] Degree of polymerization of the disaccharide repeating unit; i.e. moles of GlcN/mole of polysaccharide backbone.

[f] Estimated assuming 1 mol of linkage region D-glucuronic acid/mol polysaccharide.

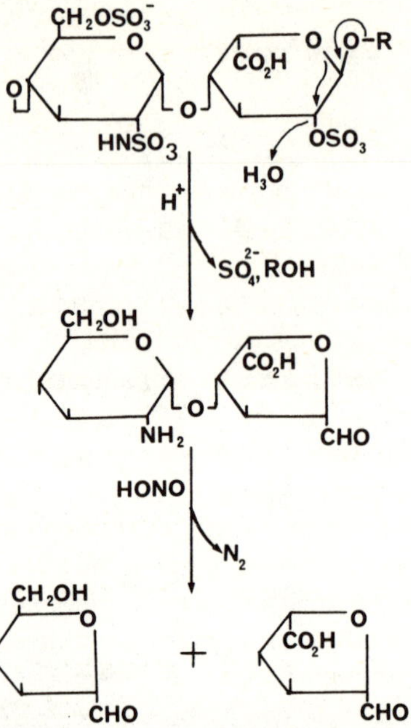

FIGURE 2.1 Mechanism proposed for the formation of the 2,5-anhydrouronic acid in the acid hydrolysis of heparin and its release as the free uronic acid in the deamination reaction. (From Shively and Conrad, 1976; reproduced with permission of publishers.)

for the anomalous behaviour of the various assays for uronic acid involving the treatment of samples with acid; these might be expected to give unique colour yields in the indole assay for anhydro-sugars and the carbazole assay for uronic acids. However, it still remains uncertain whether the heparin-like polysaccharides contain a molar excess of uronic acid residues over that of glucosamine.

2.2.3 N-acetyl-hexosamine

Studies have suggested that most if not all mammalian heparin samples contain N-acetyl groups, although some may have as little as 1 or 2% of their hexosamine residues substituted with such groups (Cifonelli and King, 1970a) (Table 2.1). Further studies by these workers (Cifonelli and King, 1972) using highly purified, beef mucosal heparin having a relatively high proportion of linkage region components (see Sections 3.5.1 and 4.3.3) was found to contain on average approximately 3 residues of N-acetyl glucosamine per chain. Identification of this component was accomplished by isolation of fragments containing this component after reaction of the polysaccharide with organic nitrites, (Section 3.3). The results indicated that the N-acetyl hexosamine residues are divided equally between the linkage region and the main part of the molecule. The residues are distributed in an isolated fashion. They also suggested that no N-sulphated glucosamine residues were involved in the linkage region. Cifonelli and King (1972) have demonstrated that most or all heparin molecules have N-acetyl glucosamine residues uniting the main protein carbohydrate linkage section to the main chain of heparin. Similar conclusions had been reported previously for hog mucosal heparin (Lindahl, 1966) and heparan sulphate (Cifonelli, 1968b; Knecht *et al.*, 1967).

Whale heparin has been shown to contain a higher proportion of N-acetyl glucosamine residues and lower sulphate content than other mammalian heparins (Yosizawa, 1964; Kotoku *et al.*, 1967) (Table 2.1). Nearly one fourth of the hexosamine in ω-heparin is N-acetyl glucosamine. It appears that such a high content of N-acetyl glucosamine residues is not unique to whale heparin as Cifonelli and King (1973) and Taylor *et al.* (1973) (Table 2.1) have found that 15% of the total glucosamine residues are N-acetylated in beef lung and hog mucosa preparations. Clam heparin (mactins) show variation of less than 2% (Cifonelli and Mathews, 1972) up to 26% content from N-acetyl glucosamine (Table 2.1). Heparan sulphates have high N-acetyl contents in the range of 40–70% (Table 2.1) and sub-fractions may be completely N-acetylated (Section 3.5.3).

2.2.4 N-unsubstituted hexosamine

While the majority of amino groups on the glucosamine residue are N-sulphated in heparin or N-acetylated in heparan sulphates, a small and variable proportion of free amino groups is also present. Evidence has been presented indicating that a small fraction, up to 0.032 in molar ratio to hexosamine for various heparin preparations (Cifonelli, 1974; Cifonelli and King, 1975), has unsubstituted amino groups. Their presence has been ascribed to losses of N-sulphate in the course of isolation of the polysaccharide or may represent a relic from an initial stage of the biosynthetic process in which N-acetyl groups are removed from the chain (Section 3.2). A sensitive method for the detection of such unsubstituted glucosamine residues is by deaminative cleavage with nitrous acid (Cifonelli, 1968ab). Less specific and even more sensitive procedures are also available for this purpose, such as coupling with 5-isothiocyanato-fluorescein to yield fluorescein heparins (Nagasawa and Uchiyama, 1978).

2.2.5 Sulphated residues

Most of the glucosamine residues in heparin carry N-sulphate groups (for review see Brimacombe and Weber, 1964), the remainder being largely N-acetylated. This is the only evidence of N-sulphate groups occurring in Nature and clearly differentiates this group of glycosaminoglycans from others.

Various methods including alkaline hydrolysis (Durand *et al.*, 1962), periodate oxidation and deamination with nitrous acid (Foster *et al.*, 1963; Wolfrom *et al.*, 1969b; Lindahl and Axelsson, 1971), methylation (Danishefsky *et al.*, 1969) and analysis of products from heparinase digestion (Dietrich, 1968; Linker and Hovingh, 1972) have established the position of the O-sulphate groups at C-6 of the hexosamine residues and at C-2 of the uronic acid residues.

The studies of Wolfrom *et al.* (1969b) and Lindahl and Axelsson (1971) have shown that D-glucuronic acid residues of heparin are not sulphated, but it is the L-iduronic acid residues that are O-sulphated at C-2. Enzymatic studies with α-L-$\Delta^{4,5}$-iduronidase on fractions from heparinase digests of whale heparin provided evidence for the presence of sulphated iduronic acid residues which was at C-2 as demonstrated by pmr spectral studies (Ototani and Yosizawa, 1974). The sulphated iduronate residue also chemically differentiates this type of glycosaminoglycan from the others. If all of the L-iduronic acid is O-sulphated at C-2 as found for heparin (Wolfrom *et al.*, 1969b; Lindahl and Axelsson, 1971) and for at least some samples of heparan sulphate (Taylor *et al.*, 1973) the data presented in Table 2.1 show that in none of the samples is there sufficient O-sulphate for all of the glucosamine residues to be O-sulphated at C-6. For whale heparin the degree of sulphation at C-6 of the glucosamine residues is very low (see also Table 2.1) (Ototani and Yosizawa, 1974; Kosakai *et al.*, 1978). Ototani *et al.* (1974) in studying digests of whale heparin with an eliminase suggested the presence of a sulphated uronic acid-N-sulphated glucosamine unit as a major repeating unit in this particular type of heparin in addition to the presence of N-acetyl glucosamine residues which might not be sulphated. That O-sulphate may vary on the N-acetyl D-glucosamine-containing sections of heparan sulphates has been reported previously for heparan sulphates from umbilical cords (Cifonelli and King, 1970b), beef lung ('heparitin sulphate A') (Dietrich and Nader, 1974) and rat brain (Margolis and Atherton, 1973). Heparitinase digestion of heparan sulphate yields N-acetylated disaccharides with O-sulphate (Hovingh and Linker, 1974).

There have been some early reports on the possibility that C-3 of the glucosamine residue may be sulphated. Evidence for this has come from alkali hydrolysis of sulphate groups from oligosaccharides obtained by deamination of N-desulphated heparin (Foster *et al.*, 1963) and from hydrolysis products of methylated, carboxyl-reduced N-acetylated heparin (Danishefsky *et al.*, 1969). However, in more recent times isolation and confirmation of the existence of disaccharides with O-sulphation at C-3 of the hexosamine residue has not been forthcoming (see Note in Added Proof. p 248.)

From a practical viewpoint, recent studies of Kosakai and Yosizawa (1979) on the labilities of the various ester sulphate groups are of interest. Treatment of heparin with 0.1M HCl at 100°C for 70 minutes removed 0.54 mol ester sulphate/mol of glucosamine besides the N-sulphate moieties. The 2-O-sulphates in L-iduronic acid were more labile (not quantitated) than 6-O-sulphates in glucosamine residues to this acid treatment. On the other hand, solvolytic desulphation with dimethylsulphoxide containing 2% pyridine at 100°C for 9 hours (see also Nagasawa *et al.*, 1977) removed 0.83 mol of ester sulphate/mol of glucosamine besides N-sulphates. In this case, 6-O-sulphates of glucosamine residues were more labile than the 2-O-sulphates of L-iduronic acid.

2.3 ANOMERIC CONFIGURATION

The fact that all the glucosaminidic and hexuronidic linkages are of the (1→4) type of heparin-like polysaccharides has been well established by chemical means (for reviews see Brimacombe and Weber, 1964; Jeanloz, 1970, 1975), whereas the anomery of such linkages has only been recently established.

2.3.1 Optical rotation

A classical characteristic of the heparin-like polysaccharides, which generally distinguishes them from other connective tissue glycosaminoglycans, is that they exhibit strongly positive optical rotatory power. Heparins give values of $[\alpha]_D$ in the range of 40°–50° and heparan sulphates have values of $[\alpha]_D$ which generally lie in the range of 60°–70° (Table 2.1). Although the strongly positive rotations are consistent with α-D-linked monosaccharides, the optical rotation values fall well below that observed for known α-D-linked polysaccharides. Values within the range of 200°–290° for $[\alpha]_D$ are found for unsulphated polysaccharides and a range of 110°–140° for $[\alpha]_D$ for their sulphated derivatives (Brimacombe and Weber, 1964). It is important to note that for both the unsulphated and sulphated polysaccharides, molar rotation $[M]_D$ values were in the range 1296°–1834° for α-D-linked polysaccharides (Brimacombe and Weber, 1964). The differences between $[\alpha]_D$ values for the α-D-linked polysaccharides and the heparin-like polysaccharides arise, in part, because of the anomery of the uronic acid residues in the latter, i.e. β-D-glucuronic and α-L-iduronic acid residues give only relatively small values of $[\alpha]_D$ (Table 2.2). (See Note, p 32.)

The differences in the optical rotatory power at 589 nm between heparin and heparan sulphate are revealed in Table 2.1. These differences appear to be related to the various proportions of L-iduronic acid and D-glucuronic acid residues in each fraction in that with higher glucuronic acid content, higher values of $[\alpha]_D$ are found. The same trend is shown for $[\alpha]_D$ values for purified disaccharide units (Table 2.2) where Kosakai and coworkers have found relatively higher values of $[\alpha]_D$ for disaccharides with glucuronic acid at the reducing end. On the basis of their values of

TABLE 2.2
Optical Rotations of Disaccharide Units
Data from Kosakai *et al.* (1978) & Kosakai & Yosizawa (1977)

Compound	Molecular weight (see Appendix 1)	$[\alpha]_D$ (deg. g^{-1})	$[M]_D$ (deg. mol^{-1})
4-O-(α-L-iduronic acid 2-sulphate)-2,5-anhydro-D-mannose (sodium salt)	462	−12.0→ −13.2	−5821
4-O-(α-L-iduronic acid 2-sulphate)-2,5-anhydro-D-mannose 6-sulphate (sodium salt)	564	−1.0→−2.0	−846
4-0(β-D-glucuronic acid) -2,5- anhydro -D- mannose 6-sulphatc (sodium salt)	462	−7.5	−3465
O-(α-D-N-acetyl glucosamine) (1→4) α-L-iduronic acid (sodium salt)	419	91.6	38380
O-(α-D-N-acetyl glucosamine) (1→4) β-D-glucuronic acid (sodium salt)	419	114.9	48143

$[\alpha]_D$ for the glucosaminosyl-uronic acid disaccharides and that for uronosyl-glucosamine disaccharides two facts emerge 1) the α-D-anomeric configuration of D-glucosamine almost totally accounts for strong positive rotation while β-D or α-L-anomers of the uronic acids have little rotatory power and 2) the values for heparin-like polysaccharides quoted in Table 2.1 are too low compared with the values of $[\alpha]_D$ near 90° for the unsulphated glucosaminosyl-uronic acid disaccharides in Table 2.2. A number of factors associated with this discrepancy may be considered. First, it is likely that in the light of current views on the formation of heparin-like polysaccharide chains *in vivo* by the action of an endoglucuronidase on the heparin-like polysaccharide proteoglycan (Section 4.3) the reducing end group of the chain will be an uronic acid residue. Otherwise, values of $[\alpha]_D$ calculated on the basis of glucosaminosyl-uronsyl values in Table 2.2 would yield an overestimate $[\alpha]_D$ for the chain. This error would be more significant for low molecular weight chains. The effects of chain conformation on the optical rotation of the heparin-like polysaccharides has received little attention (Section 4.2.3) and therefore cannot, at this stage, be taken into account in terms of an $[\alpha]_D$ prediction.

While the variation in sulphate content of heparin-like polysaccharides on their $[\alpha]_D$ has not been studied closely, it is noteworthy that sulphation of various α-D-linked polysaccharides (dextran, glycogen and nigeran were studied) did not affect their molar rotation values but markedly decreased their $[\alpha]_D$ values (Brimacombe and Weber, 1964). The effect of sulphation is also demonstrated for fractions of heparan sulphate with low sulphate and high N-acetyl contents. Fractions of heparan sulphate obtained by ion exchange methods may increase $[\alpha]_D$ to 80° for low sulphated, high N-acetylated fractions (Linker and Hovingh, 1973). Very high values of $[\alpha]_D$ in the range of 90°–130° have been obtained for two heparan sulphate fractions as obtained by gel electrophoresis procedures which were highly N-acetylated (namely 'heparin sulphate A and B', Dietrich and Nader, 1974). A value of $[\alpha]_D^{25}$ of 95° has been reported for aortic heparan sulphate (Öbrink *et al.*, 1975). The fact that the sulphate content may primarily account for differences in the $[\alpha]_D$ values of heparin and heparan sulphate fractions is also demonstrated by calculation of $[\alpha]_D$ values for the fully sulphated glucosaminosyl uronic acid disaccharide from values quoted in Table 2.2. Therefore, while N-acetylglucosamine (1→4) L-iduronic acid has an $[\alpha]_D$ of 91.6 (Table 2.2), then by assuming its sulphated derivative i.e. N-sulphated glucosamine 6-sulphate-(1→4)-L-iduronic acid 2-sulphate has the same $(M)_D$ value, its predicted value of $[\alpha]_D$ will be 58.1°. This value lies close to the experimental values of heparin and may account, in the most part, for differences in $[\alpha]_D$ values of various heparin-like polysaccharide preparations.

In the light of optical rotatory measurements as a whole, it is clear that β1,4 type linkages exist through the glucuronosyl–glucosaminosyl bond and α1,4 type linkages through the glucosaminosyl–uronosyl bond and the iduronosyl–glucosaminosyl bond. This has been confirmed recently by Hovingh and Linker (1977) in measuring $[\alpha]_D$ values on disaccharides containing $\Delta^{4,5}$-unsaturated uronic acid obtained by heparinase digestion of heparin-like polysaccharides. Other evidence for these anomeric assignments is discussed in the next section.

2.3.2 Other evidence

In the case of heparin, the presence of a N-acetyl-α-D-glucosamine (1→4) α-L-iduronic acid unit has been established by nmr spectral studies on intact heparin (Perlin *et al.*, 1970, 1972), on unsaturated disaccharides obtained by enzymatic degradation of heparin with heparinase (Perlin *et al.*, 1971; Linker and Hovingh, 1972; Ototani *et al.* 1974; Silva and Dietrich, 1975) and purified disaccharide fractions of an acid hydrolysate (Kolakai and Yosizawa, 1977).

The presence of N-acetyl D-glucosamine-(1→4)-D-glucuronic acid in heparin was suggested in early studies of Wolfrom *et al.* (1964) (see also review by Jeanloz, 1970) and more recently established from infra-red and pmr spectra of purified disaccharide fractions of acid hydrolyzates (Kosaki and Yosizawa, 1977).

The anomeric configuration of the glucuronic acid moiety had been a controversial issue. In spite of the lack of evidence for β-D-glucuronic acid residues in heparin (Wolfrom *et al.*, 1964; Silva and Dietrich, 1975) and the discrepancies associated with the action of glycuronidases on heparin oligosaccharides (Hovingh and Linker, 1977) as studied earlier (Yamagata *et al.*, 1968; Dietrich, 1969; Warnick and Linker, 1972), a large body of evidence has now accumulated to suggest that all these residues are in fact in the β-D-form. It was first shown by the liberation by β-glucuronidase of these residues located at the non-reducing end of oligosaccharides obtained by deamination of heparin (Helting and Lindahl, 1971; Linker, 1975), and subsequently by the presence of the disaccharide β-D-glucuronic acid-2,5-anhydro-D-mannose among deamination products of biosynthesized heparin of mastocytoma (Lindahl *et al.*, 1973), and of porcine and whale heparins (Kosakai *et al.*, 1978). A re-examination of a glycuronidase (isolated from *Flavobacterium*) acting on disaccharides containing $\Delta^{4,5}$-unsaturated uronic acids demonstrated that the enzyme seemed to be specific for linkage position rather than anomeric configuration and that it degraded all disaccharides derived from heparin-like polysaccharides containing $\Delta^{4,5}$-unsaturated uronic acid (Hovingh and Linker, 1977).

X-ray diffraction data on stretched films show an axial periodicity, which is consistent with the presence of β-D-glucuronic acid linkages in heparan sulphate (Atkins and Laurent, 1973). It was found that a tetrasaccharide repeat distance calculated from the diffraction patterns significantly exceeded the maximum theoretical extension of a tetrasaccharide repeat corresponding to D-glucuronic acid and N-acetyl D-glucosamine units having all α-D-(1→4)glycosidic linkages. A more plausible model was offered, involving alternating α-(1→4) D-glucosaminidic and β-(1→4)-D-glucuronidic linkages.

Finally, in viewing all the monosaccharide residues in the heparin-like polysaccharides existing in the $^1C_4$ chair form then a β-D-1,4 linkage (equivalent to α-L-1,4) through the uronosyl–glucosaminosyl bond will correspond to 1 axial:4 equatorial linkage configuration and an α-D-1,4 linkage through the glucosaminosyl–uronosyl bond will correspond to a 1 axial:4 axial linkage configuration. For iduronic acid residues that may exist in the $^4C_1$ chair form (Section 4.1) then the α-L-1,4 linkage will be 1 equatorial:4 equatorial and the α-D-1,4 linkage will be 1 axial:4 equatorial (Figure 4.1).

REFERENCES

Atkins, E. D. T. and Laurent, T. C., *Biochem. J.* **133**, 605 (1973).

Bitter, T. and Muir, H., *Anal. Biochem.* **4**, 330 (1962).

Boas, N. F., *J. Biol. Chem.* **204**, 553 (1953).

Brimacombe, J. S. and Weber, J. M., *Mucopolysaccharides. Chemical Structure, Distribution and Isolation*, Elsevier, Amsterdam (1964).

Brown, D. H., *Proc. Nat. Acad. Sci. USA* **43**, 783 (1957).

Cifonelli, J. A., in *The Chemical Physiology of Mucopolysaccharides*, G. Quintarelli (ed.), Little, Brown, Boston, Massachusetts, p 91 (1968a).

Cifonelli, J. A., *Carbohydr. Res.* **8**, 223 (1968b).

Cifonelli, J. A., *Carbohydr. Res.* **37**, 145 (1974).

Cifonelli, J. A. and Dorfman, A., *Biochem. Biophys. Res. Comm.* **7**, 41 (1962).

Cifonelli, J. A. and King, J., *Carbohydr. Res.* **12**, 391 (1970a).

Cifonelli, J. A. and King, J., *Biochim. Biophys. Acta* **215**, 273 (1970b).

Cifonelli, J. A. and King, J. *Carbohydr. Res.* **21**, 173 (1972).

Cifonelli, J. A. and King, J., *Biochim. Biophys. Acta* **320**, 331 (1973).

Cifonelli, J. A. and King, J., *Connect. Tiss. Res.* **3**, 97 (1975).

Cifonelli, J. A. and Mathews, M. B., *Connect. Tiss. Res.* **1**, 121 (1972).

Conrad, H. E., Varboncoeur, E. and James, M. E., *Anal. Biochem.* **51**, 486 (1973).
Danishefsky, I., Steiner, H., Bella, A. and Friedlander, A., *J. Biol. Chem.* **244**, 1741 (1969).
Dietrich, C. P., *Biochem. J.* **108**, 647 (1968).
Dietrich, C. P., *Biochemistry* **8**, 2089 (1969).
Dietrich, C. P. and Nader, H. B., *Biochim. Biophys. Acta* **343**, 34 (1974).
Dische, Z., J. Biol. Chem. **167**, 189 (1947).
Durand, G. J., Hendrickson, H. and Montgomery, R., *Arch. Biochem. Biophys.* **99**, 418 (1962).
Foster, A. B., Harrison, R., Inch, T. D., Stacy, M. and Webber, J. M., *J. Chem. Soc.*, p 2279 (1963).
Helting, T. and Lindahl, U., *J. Biol. Chem.* **246**, 5442 (1971).
Höök, M., Lindahl, U. and Iverius, P.-H., *Biochem. J.* **137**, 33 (1974).
Hovingh, P. and Linker, A., *Carbohydr. Res.* **37**, 181 (1974).
Hovingh, P. and Linker, A., *Biochem. J.* **165**, 287 (1977).
Jeanloz, R. W., in *The Carbohydrates*, W. Pigman, D. Horton and A. Herp (eds.), Academic Press, London and New York, Vol. 2B, p 589 (1970).
Jeanloz, R. W., in *Heparin: Structure Function and Clinical Implications*, R. A. Bradshaw and S. Wessler (eds.), Plenum Press, New York. (*Adv. Exp. Med. Biol.*, Vol. 52) p. 3 (1975).
Jorpes, J. E. and Bergström, S., *Z. Physiol. Chem.* **244**, 253 (1936).
Jorpes, J. E. and Gardell, S., *J. Biol. Chem.* **176**, 267 (1948).
Knecht, J., Cifonelli, J. A. and Dorfman, A., *J. Biol. Chem.* **242**, 6408 (1967).
Kosakai, M. and Yamauchi, F. and Yosizawa, Z., *J. Biochem.* **83**, 1567 (1978).
Kosakai, M. and Yosizawa, Z., *Carbohydr. Res.* **58**, 153 (1977).
Kosakai, M. and Yosizawa, Z., *J. Biochem.* **86**, 147 (1979).
Kotoku, T., Yosizawa, Z. and Yamauchi, F., *Arch. Biochem. Biophys.* **120**, 553 (1967).
Lindahl, U., *Biochim. Biophys. Acta* **130**, 368 (1966).
Lindahl, U. in *MTP International Review of Science. Organic Chemistry Series Two*, G. O. Aspinall (ed.), Butterworths, London and Boston, Vol. 7, Chapter 9, p. 283 (1976).
Lindahl, U. and Axelsson, O., *J. Biol. Chem.* **246**, 74 (1971).
Lindahl, U., Bäckström, G., Jansson, L. and Hallén, A., *J. Biol. Chem.* **248**, 7234 (1973).
Linker, A., *Connect. Tiss. Res.* **3**, 33 (1975).
Linker, A. and Hovingh, P., *Biochemistry* **11**, 563 (1972).
Linker, A. and Hovingh, P., *Carbohydr. Res.* **29**, 41 (1973).
Margolis, R. U. and Atherton, D. M., *Biochim. Biophys. Acta* **273**, 368 (1973).
Nagasawa, K., Inoue, Y. and Kamata, T., *Carbohydr. Res.* **58**, 47 (1977).
Nagasawa, K. and Uchiyama, H., *Biochim. Biophys. Acta.* **544**, 430 (1978).
Öbrink, B., Pertoft, H., Iverius, P.-H. and Laurent, T. C., *Connect. Tiss. Res.* **3**, 187 (1975).
Ototani, N. and Yosizawa, Z., *J. Biochem.* **76**, 545 (1974).
Ototani, N., Nakamura, K. and Yosizawa, Z., *J. Biochem.* **75**, 1283 (1974).

Perlin, A. S., Casu, B., Sanderson, G. R. and Johnson, L. F., *Can. J. Chem.* **48**, 2260 (1970).

Perlin, A. S., Mackie, D. M. and Dietrich, C. P., *Carbohydr. Res.* **18**, 185 (1971).

Perlin, A. S., Ng Ying Kin, N. M. K., Bhattacharjee, S. S. and Johnson, L. F., *Can. J. Chem.* **50**, 2437 (1972).

Rodén, L. and Horowitz, M. I., in *Glycoconjugates. Mammalian Glycoproteins Glycolipids and Proteoglycans* M. I. Horowitz and W. Pigman (eds.), Academic Press, London and New York, Vol. 2, Section 1, p. 3 (1978).

Shively, J. E. and Conrad, H. E., *Biochemistry* **15**, 3932 (1976).

Silva, M. E. and Dietrich, C. P., *J. Biol. Chem.* **250**, 6841 (1975).

Taylor, R. L. and Conrad, H. E., *Biochemistry* **11**, 1383 (1972).

Taylor, R. L., Shively, J. E., Conrad, H. E. and Cifonelli, J. A., *Biochemistry* **12**, 3633 (1973).

Warnick, C. T. and Linker, A., *Biochemistry* **11**, 568 (1972).

Wolfrom, M. L., Honda, S. and Wang, P. Y., *Carbohydr. Res.* **10**, 259 (1969a).

Wolfrom, M. L., Wang, P. Y. and Honda, S., *Carbohydr. Res.* **11**, 179 (1969b).

Wolfrom, M. L. and Rice, R. A. H., *J. Am. Chem. Soc.* **68**, 537 (1946).

Wolfrom, M. L., Vercellotti, J. R. and Horton, D. J., *J. Org. Chem.* **29**, 540 (1964).

Wolfrom, M. L., Weisblat, D. I., Karabinos, J. V., McNeely, W. H. and McLean, J., *J. Am. Chem. Soc.* **65**, 2077 (1943).

Yamagata, T., Saito, H., Habuchi, O. and Suzuki, S., *J. Biol. Chem.* **243**, 1523 (1968).

Yosizawa, Z., *Biochem. Biophys. Res. Comm.* **16**, 336 (1964).

Note: (refer p 26)

No alteration of the stereochemistry on C-1 of the hexuronic acid is involved between these two uronic acids; the change from β-D to α-L is purely one of nomenclature classification.

CHAPTER 3

Monosaccharide sequence

3.1 INTRODUCTION

On the basis of monosaccharide composition of the heparin-like polysaccharides, Figure 3.1 gives the possible oligosaccharide sequences which may occur in these polysaccharides. All possible [-uronosyl-hexosaminosyl-] and [-hexosaminosyl-uronosyl-] sequences have been represented in matrix form. It is an obvious aim of the studies on heparin-like polysaccharides to account for the arrangement of these disaccharide units in the polymer chains and to establish general rules for their arrangement. Several approaches have been employed to perform this task, namely to study 1) the mechanisms of biosynthesis of the polysaccharide chain, 2) degradation of the polysaccharides by enzymes from *Flavobacteria* and 3) chemical degradation involving nitrous acid. In the following sections of this chapter these methods will be discussed and the chapter will conclude with a summary of knowledge of particular chemical trends and sequences, which have now been established for the heparin-like polysaccharides.

3.2 POLYMER MODIFICATION DURING BIOSYNTHESIS

Although it now appears that heparin-like polysaccharides are synthesized as proteoglycans (with polysaccharide chains attached to a protein core), this section will focus primarily on polysaccharide chain synthesis and post polymer modification processes with the primary view of establishing sequence trends in the polysaccharide chain. A description of the synthesis of heparin-like

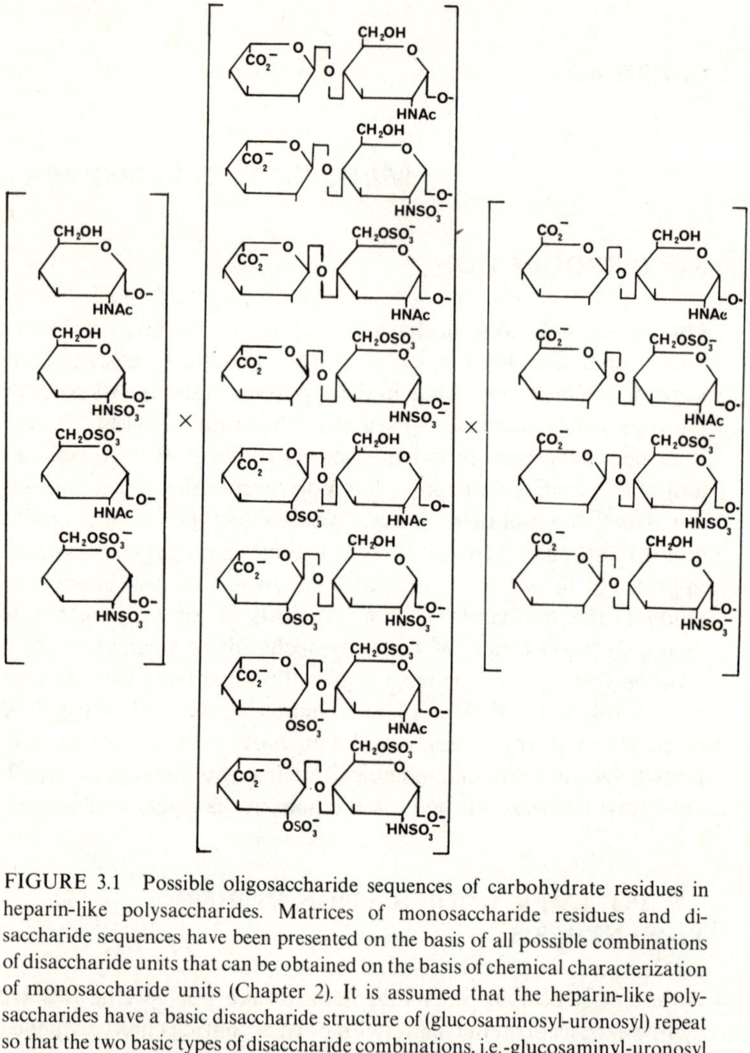

FIGURE 3.1 Possible oligosaccharide sequences of carbohydrate residues in heparin-like polysaccharides. Matrices of monosaccharide residues and disaccharide sequences have been presented on the basis of all possible combinations of disaccharide units that can be obtained on the basis of chemical characterization of monosaccharide units (Chapter 2). It is assumed that the heparin-like polysaccharides have a basic disaccharide structure of (glucosaminosyl-uronosyl) repeat so that the two basic types of disaccharide combinations, i.e.-glucosaminyl-uronosyl or uronosyl-glucosaminyl-disaccharides have been presented. Sequences with N-unsubstituted hexosamine or ester sulphate at C-3 of the hexosamine residue have not been included in the Figure.

proteoglycans or multi-chain forms is given in Section 4.3.3.

Through the elegant work of Lindahl's group in Uppsala, knowledge of the biosynthetic processes associated with heparin formation has revealed a great deal of information concerning sugar residue sequences in heparin polysaccharides. Of particular interest, in relation to sequence order of the heparin-like polysaccharides, are the following processes: 1) mechanism of N-sulphation 2) relationship between N- and O-sulphation 3) relationship between sulphation and the uronic acid C-5 epimerization reaction leading to the formation of L-iduronic acid units. Extensive and detailed reviews on heparin biosynthesis (Lindahl, 1976; Lindahl *et al.*, 1977; Rodén and Horowitz, 1978) and glycosyltransferases involved in chain formation (Rodén and Schwartz, 1975) are available. In the following sections a relatively brief account of the basic mechanisms for the formation of heparin chain will be given together with a general description of experiments, which have elucidated particular sequence formation in heparin-like polysaccharide chains. For electrostatic considerations of charge substitution in the polysaccharide chain see Footnote 3.1.

3.2.1 Chain polymerization

The principal reactions involved in the formation of heparin are now fairly well established by the use of cell-free systems derived from transplantable mouse mastocytomas. Most of the basic information was provided by the work of Silbert (1963, 1967a, b) where the polymerization process of sugar units was shown to occur by a stepwise transfer of N-acetylglucosamine and glucuronic acid from the corresponding uridine di-phosphate (UDP)-sugar precursors to the non-reducing terminal of endogenous nascent polysaccharide chains. Identical mechanisms for chain polymerization of other connective tissue glycosaminoglycans seem to operate (for review see Rodén and Schwartz, 1975) involving two types of glycosyl-transferases, namely N-acetylglucosaminyl-transferase and a glucuronosyl-transferase [11]. The latter enzyme appears quite distinct from glucuronosyl-transferase [1] involved in placement of the glucuronic acid residue to complete the polysaccharide linkage region (Section 4.3.3). In most cases, the glycosyltransferases behave in accordance with the 'one enzyme–one linkage' concept, formulated by Hagopian and Eylar (1968), which postulates that an enzyme is specific with regard to 1) the sugar transferred 2) the acceptor and 3) the position and anomeric configuration of the linkage formed. A considerable body of information is now available regarding acceptor and donor specificity in chondroitin sulphate biosynthesis, which appears to uphold the 'one enzyme–one linkage' concept. This is also probably the case for heparin biosynthesis (Helting and Lindahl, 1971, 1972). However, sequential addition of monosaccharide units to the growing polysaccharide chain during heparin biosynthesis has yet to be definitely demonstrated.

An important exception to this mode of polymerization concerns the L-iduronic acid residues of heparin, heparan sulphate and dermatan sulphate. These moieties do not appear to be formed by the incorporation of an iduronic acid residue from a nucleotide sugar into the chain, but are formed by C-5 epimerization of the glucuronic acid residues already incorporated into the growing polymer. Although this reaction appears to be unique in the biosynthesis of mammalian carbohydrates, it has a precedent in the epimerization of mannuronic acid to guluronic acid, which occurs in the course of alginic acid formation by algae and bacteria (Haug and Larsen, 1971; Larsen and Haug, 1971ab). This reaction will be further discussed below.

More recent information on chain elongation comes from Lindahl's group utilizing a microsomal fraction of a heparin-producing mastocytoma. On incubation of this fraction with UDP-[^{14}C]-GlcUA in the presence or absence of UDP-GlcNAc for varying periods of time and subsequent analysis of the [^{14}C]-labelled polysaccharides produced, by gel chromatography (Sepharose 4B), it was concluded that the polymerization of an individual polysaccharide chain may be completed within a few minutes or less. Furthermore, it was concluded that the pool of totally available primer molecules in the microsomal fraction must be large, in relation to the number of molecules engaged in simultaneous chain elongation.

3.2.2 Post polymer-modifications

It now seems well established from the work of Lindahl and collaborators that sulphation and iduronic acid formation of the heparin-like polysaccharides are post-polymer modification processes, which occur subsequent to the formation of the basic polysaccharide chain with disaccharide composition [glucuronosyl (1,4)-N-acetylglucosaminyl]. Furthermore, the modification reactions do not occur at random but in a strictly ordered manner, progressing through several distinct intermediate polymer species as catalyzed by several specific enzymes. The types of modifications finally introduced are not entirely random, in that certain sequences appear highly probable, whereas others have low probability.

The early studies of Silbert (1963) demonstrated the existance of an endogenous non-sulphated precursor of heparin, when mouse mastocytomas were incubated with UDP-GlcNAc and UDP-GlcUA. On addition of PAPS (adenosine 3′-phosphate-5′-sulphatophosphate) to the microsomal fraction containing preformed, non-sulphated polysaccharide, heparin-like polymers were obtained (Silbert, 1967ab). Further analysis of this system by Lindahl's group was made using [^{14}C]-labelled preparations isolated after incubation of a mouse mastocytoma microsomal fraction with UDP-[^{14}C]-GlcUA and UDP-GlcNAc. When PAPS was included in the incubation, containing preformed non-sulphated [^{14}C]-labelled polysaccharides, for various times, the sulphated and intermediary polysaccharides were isolated and fractionated on DEAE-cellulose (Höök *et al.*, 1975; Lindahl *et al.*, 1973). The intermediate and completed polysaccharides were then subjected to selective deamination with nitrous acid and the deamination products were analysed by gel chromatography. In using these procedures, the characterization of the range of intermediate and completed polysaccharides in terms of N-unsubstituted hexosamine, N-acetyl, N-sulphate and O-sulphate contents could be determined. This work has formed a valuable basis by which the post-polymer modifications involved in heparin formation can be studied. These experiments, together with more recent work utilizing specific microsomal enzyme assay procedures have revealed the reaction sequence outlined in Figure 3.2.

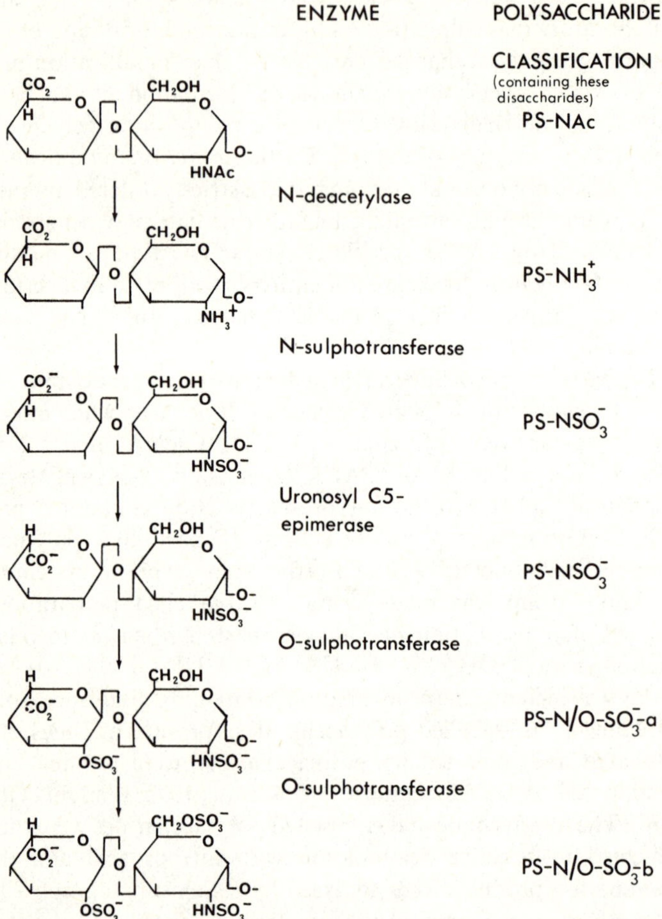

FIGURE 3.2 Main reaction pathway leading to the formation of a fully sulphated heparin disaccharide unit. There are five different polymer modification reactions required to transform the disaccharide units of the initial polymerization product, PS-NAc, into the di-O-sulphated disaccharide unit of the final product PS-N/O-SO_3^-. For cases where incomplete modification occurs or variations of this scheme, see text.

Subsequent to the polymerization of D-glucuronic acid and N-acetyl-D-glucosamine residues in alternating sequence to yield the intermediate, PS-NAc, the post-polymer modifications then commence. First, the intermediate undergoes partial N-deacetylation and is converted into a polymer, PS-NH_3^+, that contains D-glucosamine residues with unsubstituted amino groups. In the next step PS-NH_3^+ is transformed into PS-NSO_3^- which contains N-sulphated as well as N-acetylated D-glucosamine residues, D-glucuronic acid and some L-iduronic acid units but no O-sulphate groups. Finally, a major portion of the D-glucuronic acid units undergo 5-epimerization, to L-iduronic acid residues, along with the incorporation of O-sulphate groups at C-2 of L-iduronic acid and at C-6 of D-glucosamine units. The product of these final reactions, PS-N/O-SO_3^-, is a mixture of polysaccharides that includes the end-product of the overall biosynthetic process, i.e. heparin. These reactions will be discussed in more detail in the following sections.

It is to be noted that these studies have utilized the artificial system of mastocytoma-derived-microsomal fraction (105,000 g fraction) to study the cell-free biosynthesis of these heparin-like polysaccharides. It remains to be seen whether the processes *in vivo* follow the same pattern of synthesis as described.

3.2.2.1 N-SULPHATION

As described in Figure 3.2, the formation of N-sulphate groups on glucosamine involves a two step mechanism. Firstly, the N-acetylglucosamine residues are deacetylated; this reaction does not require the presence of PAPS. Further information on the deacetylase reaction has been obtained by Riesenfeld *et al.* (1979) following the development of an assay for the microsomal enzyme, which is based on the liberation of labelled acetyl groups from PS-NAc (Figure 3.2). Upon exhaustive incubation with the microsomal enzyme, only 30% to 35% of the acetyl groups were liberated from exogenous substrate used in the microsomal system. Characterization of the reaction products by deamination with nitrous acid and gel chromatography demonstrated that the N-acetyl groups are attacked in a random fashion along the polysaccharide chain. In view of the fact that most of the acetyl groups are removed in completed heparin, it would appear that endogenous polysaccharide substrate is more efficiently deacetylated. Efficient deacetylation may be dependent on the geometric organization of enzyme and substrate *in situ* together with interaction of enzymes producing N-sulphation of the deacetylated substrate.

The deacetylase reaction is of particular interest in terms of guiding subsequent polymer modification reactions. Lindahl *et al.* (1977) hypothetically view the deacetylation reaction as representing a key step for metabolic regulation. It is clear that if deacetylation does not occur, then subsequent modifications on the chain will not take place. Further, this reaction is important in terms of understanding the vectorial features of polymer modification, i.e. is the chain modified at random or do modifications take place in unidirectional manner along the chain? The sites of deacetylase attack will primarily determine the topographical placement and sequence of further biosynthetic modifications along the chain. In that random deacetylation of exogenous substrate in the microsomal system provokes interest in this context, it remains questionable as to the nature of deacetylase reaction for endogenous substrates, where maximal deacetylation occurs.

Once the deacetylated product is formed PS-NH_3^+ (Figure 3.2), the amino groups thus exposed serve as acceptors for sulphate residues in the transfer reaction with PAPS as a sulphated donor (see also Balasubramanian *et al.*, 1968). This is probably the only modification reaction that proceeds essentially to completion, as there are little, if any, free amino groups present in heparin-like polysaccharides (Section 2.2.4).

In addition to endogenous microsomal polysaccharide fractions, various exogenous preparations have been used as sulphate acceptors. Enzyme preparations from mouse mastocytoma (Eisenman *et al.*, 1967; Balasubramanian *et al.*, 1968), hen uterus (Johnson and Baker, 1973), ox lung (Foley and Baker, 1973), rat brain (George *et al.*, 1970) and arterial wall (Levy and Picard, 1976) all catalyzed the incorporation of sulphate groups from [^{35}S]-PAPS into heparin-like polysaccharides, preferentially into an N-sulphate position. N-desulphated derivatives of heparin-like polysaccharides are better sulphate acceptors than the corresponding intact polysaccharides. N-acetylated derivatives are poor acceptors.

3.2.2.2 L-IDURONIC ACID FORMATION

Originally, it was considered that the biosynthesis of L-iduronic acid units in the heparin chain was analogous to the incorporation of D-glucuronic acid units. However, it appears that an L-iduronic acid transferase is either absent or inactive in systems studied. In view of the findings of Haug and Larsen (Haug and Larsen, 1971; Larsen and Haug, 1971ab) that C-5 epimerization of preformed polymers of D-mannuronic acid residues to form L-guluronic acid may occur, the formation, by a similar mechanism, of iduronic acid units in heparin was subsequently demonstrated (Höök *et al.*, 1974a; Lindahl *et al.*, 1972).

Incubation of the mouse mastocytoma microsomal fraction with UDP-[^{14}C]-GlcUA and unlabelled UDP-GlcNAc resulted in incorporation of radioactivity into the endogenous polysaccharide. When PAPS was included in the incubate, the product polysaccharide contained L-[^{14}C]-IdUA as well as D-[^{14}C]-GlcUA. Pulse chase experiments revealed that D-[^{14}C]-GlcUA was incorporated into the polymer during the pulse period (in the absence of PAPS) and in the bound form was subsequently converted to bound L-[^{14}C] IdUA during the chase (in the presence of PAPS). Such experiments lead to the conclusion that the L-iduronic acid residues are formed by epimerization of D-glucuronic acid residues at the polymer level.

The epimerization process involves exchange of the C-5 hydrogen with hydrogen atoms of the medium (Lindahl *et al.*, 1976; Jacobsson *et al.*, 1979a). (No alteration of the stereochemistry on carbon 1 of the hexuronic acid is involved; the change from β-D to α-L is purely one of nomenclature classification.) The characteristics of the epimerase enzyme are currently being studied (Jacobsson *et al.*, 1979a). The method of assay involves measurement of tritium release from a heparin precursor polysaccharide composed of [5-^{3}H]-labelled glucuronic acid and N-sulphated glucosamine residues or the exchange of the C-5 hydrogen with

[^{3}H]-H_2O in the medium and measurement of the distribution of ^{3}H-label in deamination products.

Differential effects of exogenous polysaccharide substrates added to the microsomal system and endogenous polysaccharide substrates (formed during the course of biosynthesis in the microsomal system) have been registered for the C-5 epimerization reaction. For exogenous substrate the equilibrium reaction was characterized by an excess loss of C-5 tritium from prelabelled glucuronic acid over actual epimerization. In contrast, there was a strict proportionality between the two processes when endogenous polysaccharide was utilized by the enzyme (Jacobsson *et al.*, 1979a). These results do suggest, the none too apparent factors important in overall organization of enzyme reactions occurring in the post-polymer modification reaction scheme.

In terms of substrate specificity for the C-5 epimerase, it is clear that N-sulphated heparin-like polysaccharides are highly efficient competitors of the enzyme i.e. inhibitory activity is registered for an N-sulphated synthetic [^{3}H]-labelled substrate, whereas N-unsubstituted or N-acetylated polysaccharides are not inhibitors (Jacobsson *et al.*, 1979a). Therefore, it appears that the presence of N-sulphate groups is essential for glucuronosyl C-5-epimerization.

More recent studies by Jacobsson *et al.* (1979c) on the use of a partially N-desulphated then N-acetylated [^{14}C; 5-^{3}H] (with both 5-^{3}H and ^{14}C labelled D-glucuronic acid residues) N-sulphated intermediate (PS-NSO_3^-, Figure 3.2) with the epimerase have provided more detailed information on the role of N-substituents in relation to the epimerization reaction (Table 3.2). The modified polysaccharide was a poor substrate for the enzyme with only 20% of the ^{3}H released over a 130-hour incubation period, as compared to about 80% release with the parent polymer, PS-NSO_3^-. Analysis of this material showed that uronic acid residues located between two N-sulphated glucosamine units were essentially devoid of ^{3}H-label, as expected, whereas uronic acid residues located between two N-acetylated glucosamine units retained the label. Uronic acid units bound at C-1 to N-acetylated and at C-4 to N-sulphated glucosamine residues was a substitute for the epimerase, whereas those bound at C-1 to N-sulphated and C-4 to N-acetylated glucosamine residues were not (Table 3.2). This stipulates

TABLE 3.1
Analysis of Disaccharides in Deamination Products of Heparin-like Polysaccharides (adapted from Jacobsson *et al.*, 1979b).

Polysaccharide Preparation	Molar ratio sulphate:hexosamine	Yield of disaccharides (% of total poly-saccharide)	$IdUASO_3^-$ anh-$ManSO_3^-$	$IdUASO_3^-$ anh-Man	IdUA anh-$ManSO_3^-$	GlcUA anh-$ManSO_3^-$	IdUA anh-Man	GlcUA anh-Man
Commercial pig mucosal heparin	1.87	62	74	10	6	11	<1	<1
Whale lung heparan sulphate	0.78	24	42	35	6	12	<1	5.2
Bovine lung heparan sulphate	0.69	12	13	68	5.4	8.2	1.9	3.9

that only one out of two consecutive D-glucosamine units would have to be N-sulphated for C-5-epimerization of the interjacent D-glucuronic acid to occur. Furthermore, the glycuronosyl C-5 epimerase will not attack D-glucuronic acid residues that are adjacent to N-sulphated D-glucosamine units that are O-sulphated (Table 3.2). These experiments and other evidence have demonstrated that a sulphated L-iduronic acid cannot revert back to a D-gluco configuration nor can a uronic acid residue bound at C-1 to a 6-sulphated glucosamine unit undergo C-5 epimerization (Table 3.2). These results suggest that epimerization should then precede O-sulphation.

3.2.2.3 O-SULPHATION

There are now several lines of evidence to suggest that N-sulphation precedes O-sulphation and that iduronic acid formation is intimately related to both types of sulphation processes. The coupling of C-5-epimerization and O-sulphation was indicated early by the fact that O-sulphate groups and L-iduronic acid units appear to be accumulated within the same regions of heparin-like polysaccharides (Cifonelli and King, 1972; Taylor *et al.*, 1973).

The significance of O-sulphation in relation to the biosynthesis of iduronic acid residues was strikingly emphasized by structural analysis of the microsomal heparin-precursor polysaccharides. In this case, Höök *et al.* (1975) have shown that the polysaccharide species containing N- but no O-sulphate groups, PS-NSO_3^- (Figure 3.2), contained only small amounts of iduronic acid (15–20% of the total uronic acid). In contrast the biosynthesized O-sulphated polysaccharide derivative PS-N/O-SO_3^- contained iduronic acid constituting approximately 70% of the total uronic acid. Jacobsson and Lindahl (1980) have demonstrated recently that the proportion of D-glucuronic acid residues that undergo 5-epimerization during the conversion of PS-NSO_3^- to PS-N/O-SO_3-a (Figure 3.2) corresponds to the amount of 2-O-sulphated L-iduronic acid units in the latter fraction. On the other hand, the nonsulphated L-iduronic acid units in the final product, PS-N/O-SO_3-b (Figure 3.2) appear to be present already in PS-NSO_3^-. These findings suggest that there are two types of epimerization processes, namely one that is tightly coupled to O-sulphation of

TABLE 3.2
Substrate Specificity of Enzymes Catalysing Late Polymer-modification Reactions in Heparin Biosynthesis
(From Jacobsson and Lindahl, 1979.)

Enzyme	Polysaccharide structure[a] Accepted as substrate	 Rejected as substrate
D-glucuronosyl 5-epimerase	→GlcN→GlcUA→GlcN→ | | SO_3^- SO_3^-	→GlcNAc→GlcUA→GlcNAc→
	→GlcN→GlcUA→GlcNAc→ | SO_3^-	→GlcNAc→GlcUA→GlcN→ | SO_3^-
		6-O-SO_3^- | →GlcN→GlcUA→GlcN→ | | SO_3^- SO_3^-
L-Iduronosyl 2-O-sulpho-transferase	→IdUA→GlcN→ | SO_3^-	6-O-SO_3^- | →IdUA→GlcN→ | SO_3^-
D-glucosaminyl 6-O-sulpho-transferase	→GlcUA→GlcN→ | SO_3^-	
	→IdUA→GlcN→ | SO_3^-	
	→IdUA→GlcN→ | | 2-O-SO_3^- SO_3^-	

[a] Only structures known to be formed during biosynthesis of heparin are considered; all glycosidic linkages are thus of the appropriate 1→4 type (β-D-glucuronidic, α-L-iduronidic, α-D-glucosaminidic). The table includes some of the results described in Jacobsson *et al.* (1979a).

the iduronic acid units and one that escapes concomitant sulphation. The nature of these processes is poorly understood. However, incomplete modification such as this, which occurs infrequently for heparin, is of major importance in terms of the structural specificity of the heparin-like polysaccharides in their physiological interactions with other macromolecules. In particular, it has been proposed recently that nonsulphated L-iduronic acid in a specific location is essential for the specific interaction between heparin and antithrombin, and hence for the blood anticoagulant activity of the polysaccharide (Section 7.2.1.2.2).

There have only been limited studies performed on the properties of the sulphotransferases. Jansson *et al.* (1975) showed that preparations of N-desulphoheparin and N-acetylated heparan sulphate served as selective acceptors for incorporation of [^{35}S]-sulphate into N-sulphate and O-sulphate groups respectively on incubation with [^{35}S]-PAPS and a mouse mastocytoma microsomal fraction. These studies suggested that N- and O-sulphotransferases may be distinct enzymes.

More recently Jacobsson and Lindahl (1980) in using similar techniques, with polysaccharides containing both N-acetyl and N-sulphate groups, showed that O-sulphate groups had been incorporated only in the vicinity of N-sulphate groups. Furthermore, iduronic acid residues appeared to be sulphated only when the glucosamine unit at C-1 lacked O-sulphate at C-6 (Table 3.2). In contrast, glucosamine units could be sulphated at C-6 regardless of whether the iduronic acid unit at C-2 was sulphated or not (Table 3.2).

It was concluded that O-sulphation of iduronic acid units precedes that of the glucosamine units; O-sulphation of the glucosamine residues being the last polymer modification reaction to take place.

The fact that the presence of 6-O-sulphate groups on D-glucosamine is found to inhibit both 5-epimerization and 2-O-sulphation of adjacent uronic acid residues, is particularly interesting in relation to the relationship between heparin and heparan sulphate formation. It is conceivable that O-sulphation of the glucosamine residues may occur in early stages of the reaction sequence and this will inhibit C-5 epimerization of glucuronic acid residues and sulphation in later stages of the reaction. In this case, heparan sulphate-like regions of the molecule may be obtained. On the other hand, with the occurrence of sulphation of iduronic acid residues, heparin-like regions are obtained. The relationship between the formation of heparin and heparan sulphate will be discussed further in Section 3.5.

3.2.2.4 DEGRADATION

There is minimal information on the *in vivo* mechanisms for the degradation of heparin-like polysaccharides and how far these degradation products actually contribute to isolated heparin-like preparations. There has been an upsurge of interest in the catabolic system in view of the endoglycosidases that may act on heparin-like polysaccharides and be involved in depolymerization of macromolecular proteoglycan structures to form single polysaccharide chains (Section 4.3).

In terms of polymer-modification reactions there is evidence that heparin may be susceptible to sulphamidase activity *in vivo*. A number of studies have demonstrated the presence of sulphamidase activity *in vivo* capable of liberating inorganic sulphate from N-[S^{35}]-heparin when injected into rats (Lloyd *et al*., 1971 and earlier references). The enzyme has been isolated from various tissues including lung, liver, kidney, skin and spleen (Friedman and Arsenis, 1972, 1974). The activity appears different from the various arylsulphatases found in the same tissues. High sulphate-splitting activity on heparin has been demonstrated with mouse macrophages (Fabian *et al*., 1976).

While the catabolism of heparan sulphate is also not fully understood, studies on various genetic disorders of glycosaminoglycan metabolism (mucopolysaccharidoses) have revealed several exoenzymes capable of degrading heparan sulphate (for reviews see Section 6.1.1; also see Klein *et al*., (1978).

By the concerted action of at least four exoglycosidases (α-L-iduronidase, *β*-D-glucuronidase, α-N-acetylglucosaminidase and α-D-glucosaminidase), an α-D-glucosamine N-acetyl transferase and three sulphatases (sulphoiduronate sulphatase, glucosamine 6-sulphate sulphatase and sulphamidase) the heparan sulphate molecule could be completely degraded to monosaccharides and inorganic sulphate.

3.3 NITROUS ACID DEGRADATION

In view of the marked resistance of heparin and heparan sulphate to acid hydrolysis, structural studies on these polymers have relied heavily on cleavage of the polymers with nitrous acid to obtain oligosaccharides for further analysis. Some ambiguity has existed as far as differences in conditions of nitrous acid treatments and their specificity of attack on the heparin-like polymers are concerned (Shively and Conrad, 1976a). Shively and Conrad (1976a) have studied the reaction to establish a rationale for the differences in the reactions of heparin and amino sugars with the wide variety of nitrous acid treatments that have been employed.

Following the early demonstration that glycosides of amino sugars and heparin may be cleaved with nitrous acid whether the amino group is free or N-sulphated (Lagunoff and Warren, 1962), it has been shown that certain modifications of the deaminating reagent specifically cleave glycosides of N-sulphated amino sugars while others are specific for glycosides having a free amino group (Cifonelli, 1968; Lindahl *et al.*, 1973). All such cleavages of amino sugar glycosides convert the D-glucosamine residue primarily to an anhydro-D-mannose residue, which becomes the new reducing terminal of the oligosaccharide formed in the deamination reaction. These differences are correlated with the pH at which the deamination reaction is carried out. The actual reactant that is directly involved in the nitrosation of the amino group is not nitrous acid itself but a product NO-X formed from nitrous acid depending on the pH of the reaction mixture. At low pH values (<2.5), the nitrous acidium ion is formed

$$H^+ + HONO \rightleftarrows H_2ONO^+,$$

whereas at pH 2.5 to 4.0, for the reaction of nitrite with nitrous acidium ion, N_2O_3 is formed.

$$HONO \rightleftarrows H^+ NO_2^- \ (pK_a = 3.35)$$

$$H_2ONO^+ + NO_2^- \rightleftarrows N_2O_3 + H_2O$$

The formation of N_2O_3, a reaction catalyzed by acetate or other carboxylate ions, occurs in the pH range of 2.5 to 4.0 where both H_2ONO^+ and NO_2^- are present in relatively high concentrations.

At low pH values, H_2ONO^+ is the most likely candidate for the nitrosating reagent in the reaction for deamination of N-sulphated D-glucosamine residues, whereas N_2O_3 does not react rapidly with the N-sulphated glucosamine but does with the unsubstituted amino group at relatively high pH. With subsequent deaminative cleavage, the glycosidic bond of the D-glucosamine residue is converted to an anhydro-D-mannose residue. Specific conditions can be chosen so that only N-sulphate groups of the hexosamines react (Cifonelli, 1968; Shively and Conrad, 1976a). In this way, undegraded protein-polysaccharide linkage sections and chain segments from the interior of the molecule can be isolated from the polysaccharides, which show no loss of ester sulphate or N-acetyl groups and no cleavage of easily hydrolyzable linkages.

One problem that has arisen in the use of nitrous acid for the cleavage of heparin is that, when heparin preparations in which D-glucosamine residues are essentially 100% N-sulphated are used, the yield of disaccharides is far less than the theoretical one (Lindahl and Axelsson, 1971; Cifonelli and King, 1972). Shively and Conrad (1976a) have shown that, whereas the low pH nitrous acid treatment used to generate oligosaccharides in their study removes N--sulphate groups quantitatively, glycosidically linked D-glucosamine residues can be converted either to 2,5-anhydro-D-mannitol with glycosidic bond cleavage (deaminative cleavage reaction) or to 2-aldehydo-D-pentofuronoside without cleavage of the glucosidic bond (the deaminative ring contraction reaction). In using their low pH nitrous acid treatment on heparin for nearest neighbour analysis, the major product was, L-idosyl 2-sulphate →anhydro-D-mannitol 6-sulphate, which contained 60% of hexoses derived from the hexuronic acid residues in the original heparin and which is derived from L-iduronic acid 2-sulphate →2-N-sulphate 2-deoxy D-glucose 6-sulphate disaccharide sequences in heparin. A second product, which contained 15% of the hexoses derived from the hexuronic acid residues in the original heparin, was identified as a tetrasaccharide composed of two L-idosyl 2-sulphate residues, one anhydro-D-mannitol 6-sulphate residue (the reducing end) and an hydroxymethylpentose sulphate residue formed by deamination of a disulphated D-glucosamine residue without bond cleavage. Rapid conversion of this product to the disaccharide unit may occur in dilute acid.

These observations of Shively and Conrad (1976a) do indicate that quantitative interpretations of structural data based upon nitrous acid cleavage of heparin-like polymers must be viewed with caution.

The analysis of disaccharide products obtained by deaminative cleavage of heparin-like polysaccharides with due regard to various effects of nitrous acid on glycosidic linkages (Shively and Conrad, 1976ab) has been studied by Jacobsson *et al.* (1979b) (see Table 3.1). Apart from precautions taken in the analysis of this product with respect to the deamination treatment, a new feature in this analysis is the identification of the uronic acid components. This was achieved by means of radioactive labels selectively introduced into the uronic acid residues during biosynthesis of the polysaccharide starting material. Identification of these uronic acid compounds from deamination products of heparin by chemical, glc and pmr spectral analysis has also been made by Kosakai *et al.* (1978) and by gel and paper chromatography and the use of specific exoglycosidases (Hopwood, 1979).

The nitrous acid degradation technique has been used in combination with other degradative procedures to yield sequence information. Linker and Hovingh (1975) have subjected heparan sulphate fractions to degradation by *Flavobacterium* heparinase and the resulting oligosaccharides have been treated with nitrous acid. By employing this technique an assessment of the distribution of N-sulphated glucosamine residues in heparinase resistant oligosaccharides may be made. The results on beef lung heparan sulphates suggested that a range of fractions may be obtained where 1) at one end of the spectrum there were fractions of N-acetyl glucosamine-uronosyl-block type structures with very low degrees of N-sulphation, 2) where intermediate fractions have a predominance of sulphate substitution at the non-reducing end and low sulphation, N-acetyl glucosamine blocks at the reducing end and 3) further on the scale the molecule in general becoming more sulphate substituted.

A more recent study utilizing the nitrous acid technique has been used in combination with Smith degradation of heparan sulphate preparations from hog mucosa and beef lung (Cifonelli and King, 1977). The polysaccharides were treated with periodate to oxidize preferentially nonsulphated uronic acids and treated with acid to cleave the oxidized units. The fragments that were produced have hexosamine units at both the reducing and non-reducing ends of the oxidized fragments. An important feature of this technique is its potential for assessing the degree of sulphation of the uronic acids, their distribution along the chain and their nearest neighbours.

3.4 DEGRADATION BY ENZYMES FROM FLAVOBACTERIA

Flavobacteria, when induced to grow on heparin or heparan sulphate, produce heparinase, heparitinases, sulphatases and glucuronidases (Dietrich, 1969; Hovingh and Linker, 1970; Warnick and Linker, 1972; Dietrich *et al.*, 1973; Hovingh and Linker, 1977) which are able to degrade the heparin-like polysaccharides, when acting in concert, to their basic monosaccharide constituents.

The heparinase and heparitinases are all eliminases, which cleave the linkage between the hexosamine and uronic acid units to form an unsaturated 4,5 uronic acid which can be detected by ultra-violet absorption at 227 nm ($\varepsilon = 2800$). The eliminase action will ultimately disguise the origins of the uronic acid residue in terms of whether it was a glucuronic acid or iduronic acid. The uronic acid residue can be determined by specific isotopic labelling procedures during the biosynthesis of the polysaccharide starting material (Lindahl *et al.*, 1977; Jacobsson *et al.*, 1979b).

The heparinase appears to split the linkage between an N-sulphated or N,O-disulphated hexosamine and an iduronic acid (which may or may not require a sulphate group) (Hovingh and Linker, 1970; Silva and Dietrich, 1975; Silva *et al.*, 1976; Nader *et al.*, 1979). The enzyme will not act, if the glucosamine is N-acetylated.

Silva and Dietrich (1975) (see also Silva and Dietrich, 1974) in using commercial preparations of heparin have demonstrated that heparinase acts directly upon heparin, yielding 50–80% of trisulphated disaccharide O(-$\Delta^{4,5}$ uronic acid 2-sulphate)-(1→4)-N-sulphated glucosamine 6-sulphate (the uronic acid was inferred to be α-L-iduronic acid), the remainder being mainly tetrasaccharide units and small amounts of hexasaccharide. The yield of the trisulphated disaccharide appears to be markedly dependent on the tissue source. For mucosal heparin, the disaccharide yield is 37%, whereas for lung heparin it is 50% (Linker and Hovingh, 1972; Linker and Hovingh, 1977). Silva and Dietrich (1975) have shown that the tetrasaccharide, while resistant to repeated heparinase digestion, is in turn completely degraded by the heparitinases to form the trisulphated disaccharide and disulphated disaccharide 0(-$\Delta^{4,5}$ uronic acid)-(1→4)-N-sulphated glucosamine 6-sulphate in equal amounts (the uronic acid was inferred to be α-D-glucuronic acid). The assignment of an α-D-configuration for the glucuronic acid now appears to be incorrect (Section 2.3) and furthermore, there are no analytical data to support the assignment of glucuronic acid in the disaccharide as compared to α-L-iduronic acid. In any case, they suggest for their preparations that the unsaturated oligosaccharides, obtained by enzyme digestion, account for 95% of the original heparin molecule and the tri- and disulphated disaccharides are linked alternatively, in a proportion of 3–4:1 respectively.

Specificity assignments for heparitinase preparations have varied between several laboratories. Hovingh and Linker (1974) have purified a heparitinase which had an apparent specificity for splitting N-acetyl glucosaminyl linkages that have no neighbouring sulphate groups or only one sulphate group linked to the glucosamine residue. It will not act, when a sulphate group is linked to the uronic acid residue or when two sulphate groups are linked to the glucosamine residue.

In contrast, Silva *et al.* (1976) have been able to isolate two types of heparitinase, namely heparitinase I and heparitinase II. Heparitinase I appears to be specific for 'N-acetyl glucosamine α-(1,4)-glucuronosyl linkages' and heparitinase II appears to be specific for 'N-sulphated (or N,O-disulphated) glucosamine α-(1,4) glucuronosyl linkages'. There is now a preponderance of evidence to suggest that α-D-glucuronic acid does not occur in heparin-like polysaccharides (Section 2.3) and is thought to be β-D-glucuronic acid instead. In addition, no suitable evidence has been forthcoming in the identification of the uronic acid component in these disaccharide units released by the eliminases and that it may either be α-L-iduronic acid or β-D-glucuronic acid. With this taken into account, analysis of disaccharides obtained by the action of these enzymes on heparan sulphates cannot be fully assessed at the present time.

3.5 SEQUENCE CHARACTERISTICS

3.5.1 Linkage region

The protein-carbohydrate linkage region associated with heparin-like polysaccharides appears to have the following sequence (see Section 4.3.3):

```
|
Ser–Xyl–Gal–Gal–GlcUA
|
```

3.5.2 Linkage region/N-acetyl groups

A greater proportion of N-acetyl groups appears near the protein-carbohydrate linkage in both heparin (Cifonelli and King, 1972) and heparan sulphate (Cifonelli, 1968; Knecht *et al.*, 1967; Linker and Hovingh, 1975), whereas N-sulphate groups are predominant near the non-reducing end of the polysaccharide.

3.5.3 N-acetyl glucosamine-rich block structures

The analysis of N-acetylated regions and disaccharide units in heparin-like polysaccharides has proved difficult owing to the lack of specific degrading techniques directed at this residue. While heparitinase digests may have potential on this score, the masking of the uronic acid unit by this technique has precluded complete assessment of the possible N-acetylated disaccharides that may exist (Figure 3.1). Information about N-acetylated regions has come primarily from oligosaccharide products resistant to deamination procedures. These regions are mainly found in deamination products of heparan sulphate preparations.

Heparan sulphate has repeating units containing N-acetylglucosamine and uronic acid residues which are predominantly non-sulphated and occur largely as uninterrupted blocks (up to 15 disaccharide residues) as evidenced by both careful fractionation procedures (Linker and Hovingh, 1973) and portions of fractions resistant to heparinase and nitrous acid degradation (Cifonelli and King, 1973; Linker and Hovingh, 1975). Block structures of 10–15 disaccharide units, which include the N-acetyl block and have low sulphate contents, but do not contain heparinase sensitive linkages, have also been obtained (Linker and Hovingh, 1975) (see also Silva *et al*., 1976).

3.5.4 N-acetyl glucosamine 6-sulphate

The occurrence of O-sulphated N-acetylglucosamine residues has been shown to be associated only with unsulphated uronic acid units either at C-1 or C-4. Although its occurrence is not accounted for in post-polymer biosynthetic modification schemes (Figure 3.2), its existence has been well recognized in various preparations of heparan sulphate (Section 2.2.5).

In general, N-acetylated regions tend to have lower O-sulphate than N-sulphated areas, although N-acetylated regions are primarily sulphated at the C-6 of the glucosamine. While it has been shown that the formation of L-iduronic acid units and their O-sulphation are intimately linked, the introduction of O-sulphate in glucosamine residues may be a separate process (Jacobsson and Lindahl, 1980). The existence of 6-sulphated N-acetyl glucosamine appears to be rare in heparin preparations and is more commonly found in heparan sulphate preparations. However, its existence in heparin preparations is of particular interest, as it appears to be part of heparin binding fragment for antithrombin III (Figure 7.7). The fact that this residue inhibits C-5-epimerization and 2-O-sulphation of adjacent uronic acid residues, may deem it important in terms of the occurrence of the critically required unsulphated iduronic acid residue in position 3 of the antithrombin binding fragment of heparin (section 7.2.1.2.2). This would require O-sulphation of the glucosamine residue prior to epimerization of the glucuronic acid residue. In normal circumstances, epimerization takes place before O-sulphation as described in Figure 3.2.

3.5.5 →[GlcNAc→IdUA (or IdUASO$_{+3}^{-}$)]→ and →[IdUA (or IdUASO$_3^-$)→GlcNAc)→ sequences

These repeating units in heparin and related polysaccharides would only appear to a minor extent (Linker and Hovingh, 1975). The disaccharide of the form →[IdUA→GlcNAc]→ has been identified in the antithrombin-binding fragment of heparin (Rosenberg and Lam, 1979; Lindahl *et al.*, 1979) (see Figure 7.7). On the basis of C-5-epimerase specificity (Table 3.2) disaccharides of the form →[IdUA or (IdUASO$_3^-$)→GlcNAc]→ would not be expected to occur. However, investigations by Cifonelli and King (1975, 1977) on nitrous acid degradation of periodate-degraded fragments of heparan sulphates suggest that, although N-sulphated glucosamine is predominant at the reducing end of 2-sulphated iduronic acid, there is some evidence to indicate the existence of trisaccharides [→GlcNAc→IdUASO$_3^-$→GlcNAc→] and (→GlcNAc→IdUASO$_3^-$→GlcNSO$_3^-$]. While the occurrence of iduronic acid residues linked both at C-1 and C-4 to N-acetylglucosamine residues may suggest that C-5 epimerization of the uronic acid does not require N-sulphation of the adjacent amino sugars, Jacobsson *et al.* (1979c) have shown that the C-5 epimerase enzyme is inactive on substrates containing this sequence (Table 3.2). Resolution of these discrepancies will have to await further study.

3.5.6 →[GlcNSO$_3^-$→GlcUA]→ and →[GlcUA→GlcNSO$_3^-$] → sequences

The identification of the disaccharide →[GlcUA→GlcNSO$_3^-$]→ in various heparin-like preparations (Table 3.1) suggests that N-sulphation does not necessitate epimerization of the adjacent glucuronic acid residue bound at C-4 of the glucosamine residue.

Its occurrence as multiple sequences appears to be rare as demonstrated by analysis of deamination products of heparin-like polysaccharides (Table 3.1). In general, glucuronic acid-containing regions have lower sulphate contents as compared with iduronic acid-containing regions.

3.5.7 →[GlcNSO$_3^-$ →IdUA (or IdUASO$_3^-$)]→ and →[IdUA (or IdUASO$_3^-$)→GlcNSO$_3^-$]→ sequences

Most of the iduronic acid residues in heparin-like polysaccharides appear to be ester sulphated, whereas N-sulphated glucosamine moieties are ester sulphated in major portion only when sulphate contents are high, as in some heparin fractions (Danishefsky *et al.*, 1969; Perlin *et al.*, 1971; Taylor *et al.*, 1973 (Table 2.1); Cifonelli and King, 1977).

N-sulphated disaccharide units that contain O-sulphated iduronic acid but lack O-sulphate on the glucosamine moiety appear to prevail in ω-heparin from whale tissues (Kosakai *et al.*, 1978). Repeating units with iduronic acid in heparan sulphates are predominantly disulphated, whereas they are trisulphated repeats in heparin. Heparan sulphates contain ester sulphated iduronic acid residues almost completely in single sequences (Cifonelli and King, 1977), while heparin may have up to five or six such units in consecutive order. Relatively high sulphated heparan sulphates, such as 'heparitin sulphate 1.4' (Linker and Hovingh, 1975) have an alternating arrangement of N-acetyl and N,O-disulphated units thus conferring upon it heparitinase-resistant linkages.

Disaccharide units of the form

$$\rightarrow[\mathrm{GlcNSO_3^-}\ (\text{or}\ \underset{\substack{|\\ \text{6-O-SO}_3^-}}{\mathrm{GlcNSO_3^-}})\rightarrow\mathrm{IdUA}\ [\text{or}\ \mathrm{IdUASO_3^-}]$$

have been identified in the antithrombin-binding-fragment of heparin (see Figure 7.7). The disaccharide →[IdUA→GlcNSO$_3^-$] → would appear to occur to a minor extent, if at all, in heparin and to a minor extent in heparan sulphate preparations (Table 3.1). The disaccharide →[IdUASO$_3^-$ →GlcNSO$_3^-$]→ is more common in heparan sulphate preparations (Table 3.1), whereas the disaccharide

$$\rightarrow[\text{IdUASO}_3^- \rightarrow \underset{\displaystyle 6\text{-OSO}_3^-}{\underset{|}{\text{GlcNSO}_3^-}}]$$

is predominant in heparin preparations (Table 3.1). It is of interest that the two disaccharides which exhibit compositional invariancy when spanning preparations from heparin to heparan sulphate are the disaccharides $\rightarrow[\text{IdUA}\rightarrow\text{GlcNSO}_3^-]\rightarrow$ and $\rightarrow[\text{GlcUA} \rightarrow\text{GlcNSO}_3^-]\rightarrow$, (Table 3.1).

3.5.8 Metabolic regulation and relationships of heparin and heparan sulphate biosynthesis

In broad terms, the regulation of biosynthesis of heparin-like polysaccharides is hardly known, but in view of the large amount of work on other glycosaminoglycans, Rodén (1970) has envisaged the following control mechanisms:

1. Genetic control.
2. Regulation of enzymes involved in precursor biosynthesis as well as polymer formation (glycosyltransferases sulphotransferases, C-5 epimerase and N-deacetylase). The regulation at the enzyme level may occur in a number of different ways, including:
   a) regulation via factors commonly influencing the enzyme kinetics of one reaction,
   b) specific feedback control of certain steps in precursor formation.
3. Regulation by nutritional factors.
4. Regulation due to spatial organization of enzymes, particularly glycosyltransferases on membranes.
5. Hormonal control.
6. Regulation by interaction between the extracellular matrix and the intracellular synthetic process.

Although many of these regulation factors may conceivably operate in the synthesis of heparin-like polysaccharides (see also Footnote 3.1), there are a number of hypothetical regulation features of immediate interest. According to current concepts of glycosaminoglycan biosynthesis, the proteoglycan structure of heparin-like polysaccharides (Section 4.3) should be established by the time post-polymer modifications take place. Secondly, it is known that these modifications, like polymerization, take place extremely rapidly (Lindahl *et al.*, 1977), thus suggesting that a limited number of molecules are passed rapidly through the whole biosynthetic sequence. Thirdly, it is abundantly evident that post-polymer modification reactions do not necessarily proceed to completion (with the possible exception of N-sulphation). Incompletely modified disaccharide units occur in heparin and particularly in heparan sulphate as parts of less well-ordered regions interspersed between homo-polymeric block structures. The question of great interest is how a structure comprising protein core with pendant polysaccharide chains undergoes such biosynthetic reactions. As yet there is no information on the vectorial constraints of post-polymer modification reactions. Does chain modification proceed from one end of the chain to another or is a whole chain modified by a battery of membrane-bound enzymes acting simultaneously?

While heparin and heparan sulphate may be formed through the same biosynthetic pathway as depicted in Figure 3.2, evidence that heparan sulphate may indeed be a metabolic precursor of heparin *in vivo* is not yet forthcoming. Obviously, regulation of this pathway will determine whether the polymer will be ultimately converted to heparin or whether it will remain at a lower level of modification and consequently be classified as heparan sulphate. Lindahl *et al.* (1977) view the deacetylation reaction as representing the key step for regulation in this sequence and studies to this effect are being presently performed in their laboratory. It is clear that the degree of N-sulphation is determined by the extent of N-deacetylation. In broad terms, they view the two types of polysaccharide as being 'end products of separate, although functionally related, biosynthetic pathways'.

It is noteworthy that a number of sequence characteristics of heparan sulphates do not obey the polymer-modification scheme represented in Figure 3.2. This is particularly exemplified by the peculiar occurrence of O-sulphated N-acetylglucosamine-unsulphated uronic acid residues in heparan sulphates. In view of the studies of specificities of the C-5 epimerase and Osulphotransferase enzymes (Section 3.2.2.3) it is conceivable that O-sulphation of glucosamine residues may be an important control in determining whether a precursor polysaccharide will result in having high or low sulphated regions with corresponding high and low regions of L-iduronic acid. In this case, O-sulphation of the glucosamine residue may occur in the early stages of the reaction sequence (dependent on membrane-bound enzyme organization?) so that this will inhibit C-5 epimerization of glucuronic acid residues and their sulphation in the later stages of the reaction. When O-sulphation of the iduronic acid residues does occur, the reaction sequence is guided more or less to form heparin.

FOOTNOTE 3.1 In view of the high charge density of heparin-like polysaccharides (the line charge density may range from 1 negative charge/disaccharide for non-sulphated polysaccharides up to 4 negative charges/disaccharide unit for fully sulphated heparin), it is of interest to assess the electrostatic free energy required to place charges on such polyanions with the view of establishing any possible electrostatic limitations in terms of charge/disaccharide sequences. For example, is it possible, on electrostatic grounds, to have charge/disaccharide block type tetrasaccharide sequences in heparin of $(\text{-}4\text{-}4\text{-})_n$ or do electrostatic repulsions require charge sequences of, say, the $(\text{-}3\text{-}4\text{-})_n$ type. The following discussion of charge interactions will assume a knowledge of certain theories of polyelectrolyte behaviour, which are elaborated in Section 5.1.2.

A reasonable model of the heparin molecule in terms of micro-ion interaction is that of an infinite cylinder. The infinite cylinder models of charged polysaccharides appear to be best in describing micro-ion interaction on a theoretical basis (Manning, 1974) as applied to non heparin-like glycosaminoglycans (Preston and Snowden, 1972; Comper and Laurent, 1978) and heparin-like polysaccharides (Section 5.1.2). On the basis of the Manning (1974) theory, the problem of charge substitution for varying charge densities on a polyanion becomes trivial. In this case, it is predicted that, when the polysaccharide charge separation along the chain is less than 0.724 nm at 37°C, then monovalent counterions will condense onto the excess polysaccharide charge and inactivate it until the effective line charge separation becomes 0.724 nm. Therefore, for all polyanions with charge separations less than 0.724 nm. Therefore, for all polyanions with charge separations less then 0.724 nm, counterion condensation and masking of polyanion charge will result in all polyanions having the same effective charge, and resultant effective electrostatic field generated by 'free' polyanion charge, irrespective of their total charge. In that the critical charge separation of 0.724 nm is approximately satisfied for the non-sulphated heparin-like polysaccharides, it is viewed that sulphation of this polysaccharide does not undergo electrostatic restriction in terms of the stages of its sulphation. All the intermediates in this reaction sequence will have the same effective charge. The only problem in charging these units is electrostatic free energy required to place a charge on a polyanion with 0.724 nm effective charge spacing.

The work, W_{el}, performed in charging a length of the cylinder, L, and of radius, r_c, is given by Hill (1955) as

$$W_{el}=\frac{Z^2e^2}{DL}\left[\frac{K_0(\kappa a_m)}{\kappa a_m K_1(a_m)}+\ln\frac{a_m}{r_c}\right]\times\frac{N}{4.186\times 10^{10}} \quad 3.1$$

where Z is the charge spread uniformly over the cylinder surface, e is the elementary charge ($=4.803\times 10^{-10}$ e.s.u), $K_0(x)$ and $K_1(x)$ represent modified Bessel functions of the second kind (Abramowitz and Stegun, 1965) D is the dielectric constant, and a_m is the distance with which the centre of a micro-ion can approach the centre of the cylinder such that $a_m > r_c$ and κ is the Debye–Hückel parameter given by the equation

$$\kappa=\left(\frac{8II\ Ne^2}{1000\ DkT}\right)^{1/2}(I)^{1/2} \quad 3.2$$

where k is Boltzmann's constant, T is the temperature ($=310°$K) and I represents the ionic strength and is expressed as

$$I=1/2\left(\sum_i C_i Z_i^2\right)$$

where C_i is the molar concentration of any mobile ion and Z_i its charge; the summation extending over all kinds of mobile ions present.

Experiments on glycosaminoglycan synthesis would suggest that the channels of the rough and smooth endoplasmic reticulum and vesicles of the Golgi apparatus have been implicated as sites of polysaccharide polymerization and post-polymer modification reactions (for reviews Rodén, 1970; Jamieson and Palade, 1977). These enzymes involved in heparin-like polysaccharide synthesis appear to be firmly bound to subcellular membranes (Helting and Lindahl, 1972; Helting, 1972; Jacobsson *et al.*, 1979a). The existence of glycosyltransferases on the cell surface (Roseman, 1970) is of particular interest with regard to the biosynthesis of heparan sulphate — a known cell surface component.

In that the location of the biosynthetic reactions is known in broad terms, the local environment and organization is not. In calculation of W_{el}, by assuming a dielectric constant $D=74$ at 37°C (as in water) will minimize the calculated electrostatic work done in charging the polyion. If such a process occurs in a lipid environment, then the value of D will be in the range of 2–10 and will effectively increase W_{el}, as this parameter is proportional to $D^{-1/2}$. If the charge substitutions occur in an aqueous environment with $D=74$ then the ionic strength will be taken as physiological and $I=0.15$ so the value of κ from equation 3.2 is 1.23×10^7.

In terms of the charging process, the sulphation mechanism seems most appropriate, theoretically, as it occurs within the interior regions of the chain. Calculation of the electrostatic work necessary to bring a quaternary charged molecule (PAPS), from infinity to the interior regions of the heparin-like polysaccharide yields a value 9.4 Kcal/mol (where Z initially is taken as 5 charges/0.724 nm and finally as 1 charge/0.724 nm subsequent to the charging process). Given the approximate nature of the model, the lack of kinetic and thermodynamic analysis on the charging process *in situ*, it could be stated at this time that the electrostatic contribution to the free energy of polymer-sulphation is significant.

REFERENCES

Abramowitz, M. and Stegun, I. A. (eds.), *Handbook of Mathematical Functions*. National Bureau of Standards. *Applied Mathematics Series* 55 (1968).

Balasubramanian, A. S., Joun, N. S. and Marx, W., *Arch. Biochem. Biophys.* **128**, 623 (1968).

Cifonelli, J. A., *Carbohydr. Res.* **8**, 233 (1968).

Cifonelli, J. A. and King, J., *Carbohydr. Res.* **21**, 173 (1972).

Cifonelli, J. A. and King, J., *Biochim. Biophys. Acta* **320**, 331 (1973).

Cifonelli, J. A. and King, J., *Connect. Tiss. Res.* **3**, 97 (1975).

Cifonelli, J. A. and King, J., *Biochemistry* **16**, 2137 (1977).

Comper, W. D. and Laurent, T. C., *Physiol. Rev.* **58**, 255 (1978).

Danishefsky, I., Steiner, H., Bella, A. and Friedlander, A., *J. Biol. Chem.* **244**, 1741 (1969).

Dietrich, C. P., *Biochemistry* **8**, 2089 (1969).

Dietrich, C. P. and Nader, H. B., *Biochim. Biophys. Acta* **343**, 34 (1974).

Dietrich, C. P., Silva, M. E. and Michelacci, Y. M., *J. Biol. Chem.* **248**, 6408 (1973).

Eisenman, R. A., Balasubramanian, A. S. and Marx, W., *Arch. Biochem. Biophys.* **119**, 387 (1967).

Fabian, I., Bleiberg, I. and Aronson, M., *Biochim. Biophys. Acta* **437**, 122 (1976).

Friedman, Y. and Arsenis, C., *Biochem. Biophys. Res. Commun.* **48**, 1133 (1972).

Friedman, Y. and Arsenis, C., *Biochem. J.* **139**, 699 (1974).

Foley, T. and Baker, J. R., *Biochem. J.* **135**, 187 (1973).

George, E., Singh, M. and Bachhawat, B. K., *J. Neurochem.* **17**, 189 (1970).

Hagopian, A. and Eylar, E. H., *Arch. Biochem. Biophys.* **126**, 785 (1968).

Haug, A. and Larsen, B., *Carbohydr. Res.* **17**, 297 (1971).

Helting, T., *J. Biol. Chem.* **247**, 4327 (1972).

Helting, T. and Lindahl, U., *Acta. Chem. Scand.* **26**, 3515 (1972).

Hill, T. L., *Arch. Biochem. Biophys.* **57**, 229 (1955).

Höök, M., Lindahl, U., Bäckström, H., Malmström, A. and Fransson, L.-Å., *J. Biol. Chem.* **249**, 3908 (1974a).

Höök, M., Lindahl, U. and Iverius, P.-H., *Biochem. J.* **137**, 33 (1974b).

Höök, M., Lindahl, U., Hallén, A. and Bäckström, G., *J. Biol. Chem.* **250**, 6065 (1975).

Hopwood, J. J., *Carbohydr. Res.* **69**, 203 (1979).

Hovingh, P. and Linker, A., *J. Biol. Chem.* **245**, 6170 (1970).

Hovingh, P. and Linker, A., *Carbohydr. Res.* **37**, 181 (1974).

Hovingh, P. and Linker, A., *Biochem. J.* **165**, 287 (1977).

Jacobsson, I., Bäckström, G., Höök, M., Lindahl, U., Feingold, D. S., Malmström, A. and Rodén, L., *J. Biol. Chem.* **254**, 2975 (1979a).

Jacobsson, I., Höök, M., Petersson, I., Lindahl, U., Larm, O., Wirén, E. and von Figura, K., *Biochem. J.* **179**, 77 (1979b).

Jacobsson, I. and Lindahl, U., *J. Biol. Chem.* **255**, 5094 (1980).

Jacobsson, I., Lindahl, U., Jensen, J., Rodén, L., Prihar, H. and Feingold, D. S., (manuscript in preparation).

Jansson, L., Höök, M., Wasteson, Å. and Lindahl, U., *Biochem. J.* **149**, 49 (1975).

Johnson, A. H. and Baker, J. R., *Biochim. Biophys. Acta* **320**, 341 (1973).

Klein, U., Kresse, H. and von Figura, K., *Proc. Natl. Acad. Sci. USA* **75**, 5185 (1978).

Knecht, J., Cifonelli, J. A. and Dorfman, A., *J. Biol. Chem.* **242**, 6408 (1967).

Kosakai, M., Yamauchi, F. and Yosizawa, Z., *J. Biochem.* **83**, 1567 (1978).

Lagunoff, D. and Warren, G., *Arch. Biochem. Biophys.* **99**, 396 (1962).

Larsen, B. and Haug, A., *Carbohydr. Res.* **20**, 225 (1971a).

Larsen, B. and Haug, A., *Carbohydr. Res.* **17**, 287 (1971b).

Levy, P. and Picard, J., *Eur. J. Biochem.* **61**, 613 (1976).

Lindahl, U., in *MTP International Review of Science. Organic Chemistry Series Two*, G. O. Aspinall (ed.), Butterworths, London and Boston, Vol. 7, Chapter 9, p. 283 (1976).

Lindahl, U. and Axelsson, O., *J. Biol. Chem.* **246**, 74 (1971).

Lindahl, U., Bäckström, G., Höök, M., Thunberg, L., Fransson, L.-Å. and Linker, A., *Proc. Natl. Acad. Sci. USA* **76**, 3198 (1979).

Lindahl, U., Bäckström, G., Jansson, L. and Hallén, A., *J. Biol. Chem.* **248**, 7234 (1973).

Lindahl, U., Bäckström, G., Malström, A. and Fransson, L.-Å., *Biochem. Biophys. Res. Commun.* **46**, 985 (1972).

Lindahl, U., Höök, M., Bäckström, G., Jacobsson, I., Riesenfeld, J., Malström, A., Rodén, L. and Feingold, D. S., *Fed. Proc.* **36**, 19 (1977).

Lindahl, U., Jacobsson, I., Höök, M., Bäckström, G. and Feingold, D. S., *Biochem. Biophys. Res. Commun.* **70**, 492 (1976).

Linker, A. and Hovingh, P., *Biochemistry* **11**, 563 (1972).

Linker, A. and Hovingh, P., *Carbohydr. Res.* **29**, 41 (1973).

Linker, A. and Hovingh, P., *Biochim. Biophys. Acta* **385**, 324 (1975).

Linker, A. and Hovingh, P., *Fed. Proc.* **36**, 43 (1977).

Lloyd, A. G., Embery, G. and Fowler, L. J., *Biochem. Pharmacol.* **20**, 637 (1971).

Manning, G., in *Charged and Reactive Polymers. I. Polyelectrolytes.* E. Selegny (ed.), D. Reidel Publishing Company, Dordrecht–Holland, p. 9 (1974).

Nader, H. B., Cohen, D. M. and Dietrich, C. P., *Biochim. Biophys. Acta.* **582**, 33 (1979).

Perlin, A. S., Mackie, D. M. and Dietrich, C. P., *Carbohydr. Res.* **18**, 185 (1971).

Preston, B. N. and Snowden, J., *Biopolymers* **11**, 1627 (1972).

Riesenfeld, J., Höök, M. and Lindahl, U. J., *Biol. Chem.* (in press).

Rodén, L., in *Metabolic Conjugation and Metabolic Hydrolysis*, W. H. Fishman (ed.), Academic Press, New York, Vol. 2, p. 346 (1970).

Rodén, L. and Horowitz, M. I., in *Glycoconjugates. Mammalian Glycoproteins, Glycolipids and Proteoglycans*, M. I. Horowitz and W. Pigman (eds.),

Academic Press, London and New York, Vol. 2, Section 1, p. 3 (1978).
Rodén, L. and Schwartz, N. B., in *MTP International Review of Science. Biochemistry of Carbohydrates*, W. J. Whelan (ed.), University Park Press, Baltimore, Vol. 5, p. 95 (1975).
Rosenberg, R. D. and Lam, L., *Proc. Natl. Acad. Sci. USA* **76**, 1218 (1979).
Shively, J. E. and Conrad, H. E., *Biochemistry* **15**, 3932 (1976a).
Shively, J. E. and Conrad, H. E., *Biochemistry* **15**, 3943 (1976b).
Silbert, J. E., *J. Biol. Chem.* **238**, 3542 (1963).
Silbert, J. E., *J. Biol. Chem.* **242**, 2301 (1967a).
Silbert, J. E., *J. Biol. Chem.* **242**, 5146 (1967b).
Silva, M. E. and Dietrich, C. P., *Biochem. Biophys. Res. Commun.* **56**, 965 (1974).
Silva, M. E. and Dietrich, C. P., *J. Biol. Chem.* **250**, 6841 (1975).
Silva, M. E., Dietrich, C. P. and Nader, H. B., *Biochim. Biophys. Acta* **437**, 129 (1976).
Taylor, R. L., Shively, J. E., Conrad, H. E. and Cifonelli, J. A., *Biochemistry* **12**, 3633 (1973).
Telser, A., Robinson, H. C., and Dorfman, A., *Proc. Nat. Acad. Sci.* **54**, 912 (1965).
Warnick, C. T. and Linker, A., *Biochemistry* **11**, 568 (1972).

CHAPTER 4

Secondary and higher order structures

4.1 RESIDUE CONFORMATION

It is clear that the heparin-like polysaccharides exhibit conformational dependence on the environmental conditions in which they are studied; such conditions include ionic strength, pH, counter-ion type, complexes with polycations and in condensed phases. In view of the primary dependence of conformation on the electrostatic potential environment of the polysaccharide chain, intrinsic parameters of the chain are of importance e.g. molecular weight, charge substitution and disaccharide sequence. Therefore, conclusions drawn from particular conformational states observed under particular conditions for particular preparations should be considered carefully in view of relating these results to conformational states that may occur in physiological situations.

Comprehensive studies on conformational states of heparin-like polysaccharides ranging from monosaccharide, oligosaccharide to polysaccharide levels have yet to be performed.

4.1.1 X-ray diffraction

Although it has been known for many years that barium-heparin can form crystals (Charles and Scott, 1936; Wolfrom *et al.*, 1943; for review see Kavanagh and Jaques, 1974) these have not proved suitable for X-ray diffraction analysis. Definitive work on X-ray diffraction patterns of glycosaminoglycans began in 1970 by Atkins and Sheehan (1972) when they utilized ordered fibres of hyaluronate in stretched films. This was a process developed earlier by Sylvén and Ambrose (1955) and Bettelheim (1958, 1959).

Before proceeding to a description of X-ray diffraction studies on heparin-like polysaccharides it is important to appreciate some of the well documented limitations of the technique. In terms of orienting glycosaminoglycan chains in films it is clear than 100% crystal formation or perfectly ordered films are never obtained. Normally a parallel alignment of particular segments, exhibiting regular primary sequence structure, to form a coherent domain of limited extent in a direction perpendicular to the molecular chain axis is obtained. The linear polysaccharide chains usually can be oriented further, and sometimes also crystallized by application of constant tension, at an appropriate temperature and relative humidity (Arnott and Winter, 1977). It is to be emphasized that the X-ray diffraction technique gives no information about segments of the molecule, or those molecules which do not align or crystallize. It is generally considered that the repeat units evident from X-ray diffraction data correspond to certain block structures selected by their ability to crystallize. The molecular weight dependence of the heparin-like polysaccharides on their ability to align is as yet unknown.

A further objection often directed towards the study of these structures in solid-state is that their conformation may be different from that in solution owing to the relatively greater degrees of conformational freedom that the molecule may take in solution. It is probable that these objections are quite valid when considering the apparently different conformations these polysaccharides may assume in solution (Section 4.2). However, the importance of the environmental state in which these molecules are studied is in relation to their physiological environment *in vivo*. In this case, it is evident that heparin-like polysaccharides in solution with simple counterion (e.g. Na^+, or Ca^{2+}) existing *in vivo* have yet to be found and that due to their highly anionic nature they are normally occupied in complexes with other macroions both in solution and in the solid state (Chapters 6 and 7). In view of these considerations, the conformational aspects of the heparin-like polysaccharides in condensed phases are equally as important as their conformations in solution.

4.1.1.1 HEPARIN

X-ray fibre diffraction patterns have been obtained from pig mucosal heparin (Nieduszynski and Atkins, 1973) and rat skin macromolecular heparin (Atkins *et al.*, 1974). The values of axially projected periodicities lie in a relatively broad range of 0.8–0.87 nm (Atkins and Nieduszynski, 1977) and have been shown to depend on the counterion type and relative humidity (Atkins *et al.*, 1974). The molecular shape for the heparin chain has been basically represented by a 2-fold helix having a tetrasaccharide repeat unit. The sodium salt of heparin crystallizes in a triclinic unit cell with dimensions, $a = 1.102$ nm, $b = 1.201$ nm, c (fibre axis) $= 1.680$ nm, $\alpha = 90.0°$, $\beta = 116.2°$ and $\gamma = 107.3°$ (Nieduszynski *et al.*, 1977; Atkins and Nieduszynski, 1977). The distribution of water and counterions in this unit cell has not been clearly defined.

The interpretation of the residue conformation from X-ray diffraction data has been subject to varying conclusions. This has been primarily due to the interpretation of the range of values of the axially projected disaccharide repeat distances (0.8–0.87 nm). On one hand, Atkins and coworkers view these differences as arising in the extension of the $^4C_1$ and $^1C_4$ chair forms of the L-iduronic acid residue (Figure 4.1). This residue may display conformational flexibility depending on its state of sulphation and its environment. (Their interpretations of molecular conformations have been based on the assumption that the diffracting unit represents a highly O-sulphated region in the heparin-molecule with alternating N-sulphated glucosamine (in the $^4C_1$ chair form) and iduronic acid residues.) A further possible conformation of the sulphated iduronate residue in heparin, which has been considered, is the skew boat conformation $^1S_3$ (Figure 4.1) (Atkins and Nieduszynski, 1977; Nieduszynski *et al.*, 1977). The general conclusion from their work is that the sulphated iduronate residue may exist in the $^1C_4$ or $^1S_3$ residue conformation; they were unable to distinguish between these two conformations. On the other hand, Nagarajan and Rao (1979) have refuted these interpretations. They suggest an alternative model for the tetrasaccharide unit in which the glucosamine and one of the uronic acids exists in the $^4C_1$ conformation and the other uronic acid (probably sulphated) in the $^1C_4$ conformation. They suggest that the range of axial periodicities observed for heparin are not a result of conformational flexibility of the

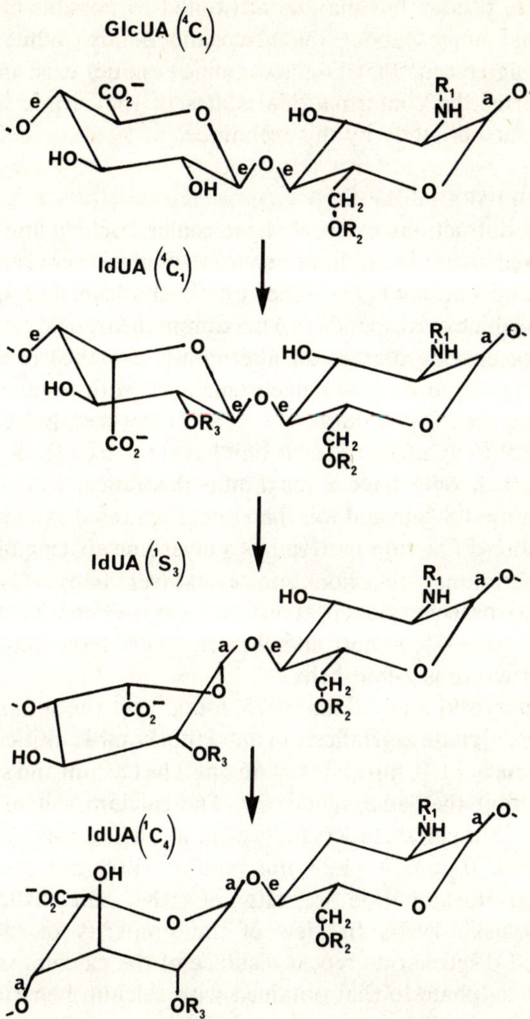

FIGURE 4.1 Various conformational forms for the uronosyl-glucosaminosyl disaccharide unit in heparin-like posaccharides. The glucosamine unit takes on the $^4C_1$ chair conformation and $R_1 = Ac$, H or SO_3^-, R_2 and $R_3 = H$ or SO_3^-. The iduronic acid residue may exhibit some conformational flexibility as described in text.

iduronate residue but may be attributed to possible changes in rotational angles about the glycosidic bonds. While there is general agreement that D-glucosamine residues exist in the $^4C_1$ chair form, the conformational states of the uronic acids will require further study by this technique.

4.1.1.2 HEPARAN SULPHATE

X-ray diffractions of ordered molecular conformations in the condensed phase for sodium heparan sulphate were reported by Atkins and Laurent (1973). They obtained a layer-line spacing of 1.86 nm which corresponds to a maximum theoretical extension of a tetrasaccharide unit of an alternating sequence of α-(1,4)-D-glucosamine and β-(1,4)-D-glucuronic acid, both residues existing in the energetically favourable $^4C_1$ chair form (see also Arnott and Winter, 1977). A tetrasaccharide unit having all α-D-(1,4)-glycosidic linkages can only have a maximum theoretical extension of 4×0.45 nm $= 1.80$ nm and was therefore discounted as a model. The X-ray fibre-diffraction pattern has a layer-line spacing of 1.86 nm with meridional reflexions on even-layer lines. The axially projected disaccharide repeat distance was 0.93 nm (Nieduszynski and Atkins, 1975; Arnott and Winter, 1977), indicating that the basic structure is 2-fold helix.

Nieduszynski and Atkins (1975) found that the sodium salt of heparan sulphate crystallizes in an orthorhombic unit-cell with $a = 1.18$ nm, $b = 1.10$ nm and $c = 1.86$ nm. The barium and strontium salts exhibit the same value of c. The calcium salt of heparan sulphate (human aorta) crystallizes in an orthorhombic unit cell with $a = 1.70$ nm, $b = 1.27$ nm, and $c = 1.68$ nm and axially projected disaccharide repeats of 0.84 nm (Atkins and Nieduszynski, 1976). In view of the similarity of the axially projected disaccharide repeat distance of the calcium salt of the heparan sulphate to that obtained with calcium heparin, Atkins and Nieduszynski have suggested that the preparation is a mixture of a 'heparan sulphate-like' and a 'heparin-like' phase and that the former crystallizes preferentially in the sodium salt form (Atkins and Nieduszynski, 1977).

Elloway and Atkins (1977) have generated possible helical models consistent with the X-ray diffraction results for sodium

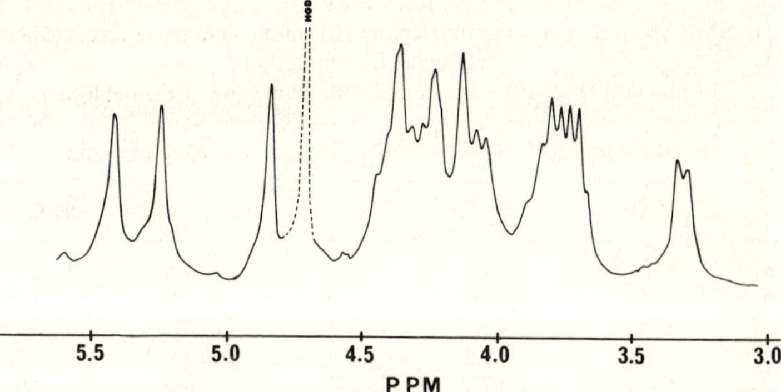

FIGURE 4.2 ^{1}H-NMR spectrum of heparin, at 270 MHz, 35°C (from Gatti *et al.*, 1978; reproduced with permission of publishers).

heparan sulphate. Model building, utilizing backbone and side group coordinates to generate conformational maps and conformationally allowed regions, was performed for both residues in the $^4C_1$ chair form. For the linkage, β-D-GlcUA-(1→4)-α-D-GlcNAc (1 eq–4 eq) only one hydrogen bond is possible, i.e. GlcUA–O(5)... H–O(3)–GlcNAc with O(3) acting as donor (see also Arnott and Winter, 1977). For the linkage α-D-GlcNAc-(1→4)-β-D-GlcUA again only one hydrogen bond is possible in the conformationally allowed, i.e. GlcNAc–N–H... O(3)–GlcUA with N acting as donor.

4.1.2 ^{1}H-nmr (pmr) spectroscopy

Early studies of ^{1}H-nmr spectroscopy of heparin at 220 MHz in D_2O solution (Perlin *et al.*, 1970) suggested that the glucosamine and iduronate residues have the $^4C_1$ (D) and $^1C_4$ (L) conformation respectively. A full analysis of the ^{1}H-nmr spectrum of the major residues of heparin (commercial beef lung) at 35°C and 90°C has recently been performed by Gatti *et al.* (1978) and is shown in Figure 4.2. The chemical shifts and coupling constants for the ring protons are given in Table 4.1. The coupling constants found for the ring protons of the glucosamine residues confirm the expected $^4C_1$ (D) conformation. J_{12} is small (3.66 Hz) in accord with a gauche relationship between H-1 and H-2, whereas the other

TABLE 4.1
^{1}H-NMR Spectral Parameters for Heparin [Chemical Shifts (δ_i) in ppm; coupling constants (J_{ik}) in Hz]
(From Gatti *et al.*, 1978 reproduced with permission of the publisher)

	Iduronic Acid Unit			Glucosamine Unit	
	35°C	90°C		35°C	90°C
δ_1	5.225	5.218	δ_1	5.401	5.369
δ_2	4.347	4.350	δ_2	3.284	3.298
δ_3	4.205	4.213	δ_3	3.673	3.691
δ_4	4.106	4.120	δ_4	3.771	3.750
δ_5	4.820	4.772	δ_5	4.035	4.049
J_{12}	2.64	3.29	δ_6	4.407	4.378
J_{23}	5.90	6.10	$\delta_{6'}$	4.278	4.284
J_{34}	3.44	3.60	J_{12}	3.66	3.57
J_{45}	3.09	3.14	J_{23}	9.98	9.88
			J_{34}	9.09	8.91
			J_{45}	9.23	9.23
			J_{56}	2.92	2.92
			$J_{56'}$	2.15	2.15
			$J_{66'}$	−11.23	−11.23

constants are large (9–10 Hz) as required for *trans*-axial orientations (Figure 4.3). The small values of J_{56} and $J_{56'}$ indicate the conformer (e) in Figure 4.3 is preponderant, where minimization of electrostatic and steric interactions of the $-OSO_3^-$ group probably occur.

By contrast, the coupling constants for protons of the iduronic acid residues are all <6 Hz, which rules out the $^4C_1$ (L) conformation. In the alternate $^1C_4$ (L) conformation, as shown by Newman projections, the vicinal protons (all *gauche*) occur in three environments with respect to their antiperiplanar atoms (Figure 4.3): a) H-1/H-2 are opposite to a carbon and an oxygen; b) H-2/H-3 and also c) H-3/H-4, to two carbon atoms: and d) H-4/H-5, to two oxygen atoms. On the basis of this model the experimental values obtained are all in reasonably good agreement with theory. The coupling constants are essentially temperature independent, which strongly suggests that the $^1C_4$(L) conformation of the iduronic acid as well as the $^4C_1$(D) conformation of the

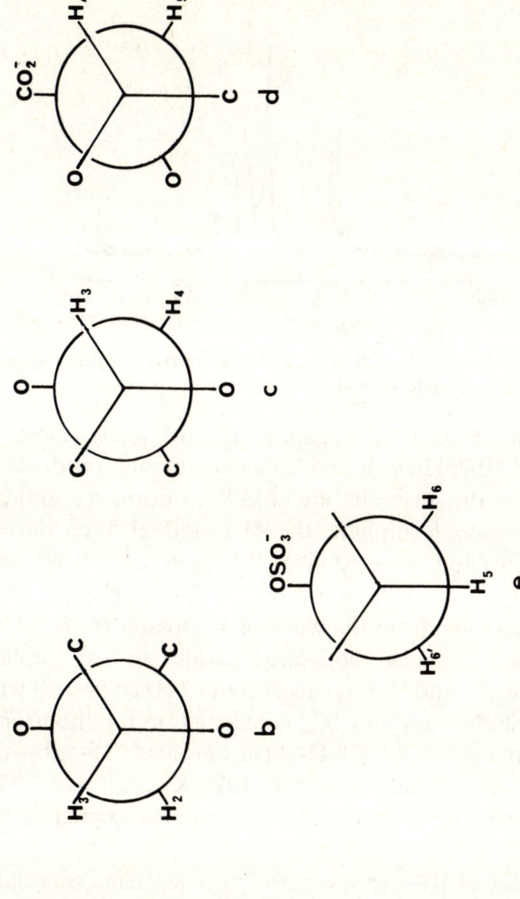

FIGURE 4.3 Newman projections for ring protons of iduronic acid (a, b, c, d) and protons at C-6 of glucosamine (e) (adapted from Gatti *et al.*, 1978; reproduced with permission of the publisher).

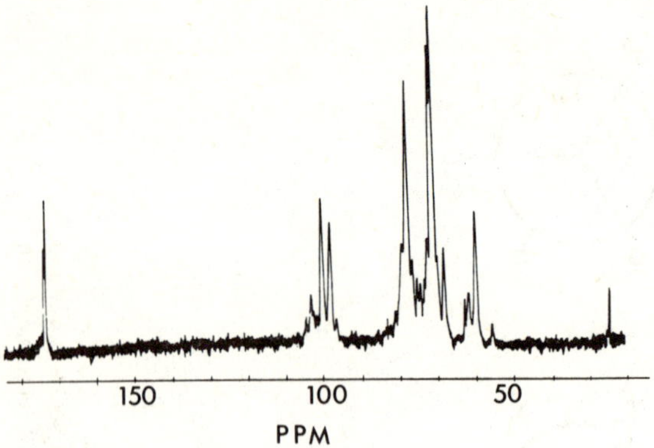

FIGURE 4.4 ^{13}C-NMR spectra (20 MHz, D_2O) of heparin (source unknown) (from Casu *et al.*, 1979; reproduced with permission of the publishers).

aminosugar residue are readily maintained and highly stable.

Kosakai *et al.* (1978) have found from deamination products of heparin that in the disaccharide 4-O-(α-L-iduronic acid)-2,5 anhydro-D-mannose 6-sulphate the H-1 and H-5 of the L-isuronic acid residue were shifted 0.3 ppm and 0.07 ppm respectively upfield as compared to the 2-O-sulphated acid. Similar data has come from the work of Jacobsson *et al.* (1979). According to Jacobsson *et al.*, both disaccharides showed coupling constants for the H-1 and H-2 signals of about 2 Hz consistent with the α-L-*ido* configuration and $^1C_4$ conformation for the uronic acid residues. Analysis of 4-O-(β-D-glucuronic acid)-2,5-anhydro-D-mannose 6-sulphate (Jacobsson *et al.*, 1979; Kosakai *et al.*, 1978) demonstrated that the D-glucuronic acid residue existed as the $^4C_1$ conformer.

Another feature of ^1H-nmr spectroscopy of heparins, especially distinguishing those preparations that contain appreciable amounts of N-acetyl glucosamine residues and those containing only a trace proportion of such residues, is the readily detectable methyl proton signal (minor peak at 2.1 ppm not shown in Figure 4.2).

4.1.3 ^{13}C-nmr spectroscopy

The ^{13}C-nmr spectrum of heparin is shown in Figure 4.4 and values of the major ^{13}C-assignments are given in Table 4.2. The

TABLE 4.2
Major ^{13}C-NMR Spectral Parameters for Heparin [IdUA/GlcUA molar ratio 3:1]
(Numbers in brackets are J_1(CH) coupling constants)
(From Fransson *et al.*, 1978)

Anomeric resonances	
102.6	GlcUA → GlcN
99.6 (173)	IdUASO_3^- → GlcN
97.7 (174)	GlcNSO_3^- → IdUASO_3^-
Other assignable resonances	
175.5	carboxyls
67.2	GlcN C(6)-SO_4^-
60.5	GlcN C(6)-OH
58.5	NSO_3^- } Hexosamine C(2)
54.1	NAc } Hexosamine C(2)
22.6	Acetyl CH_3

values in brackets of Table 4.2 are the J_1(CH) coupling constants (173–174 Hz) and are in accord with an axial orientation for O–1 in both the iduronic acid and glucosamine residues (Perlin, 1977). This is consistent with a $^1C_4$ chair form for iduronic acid and a $^4C_1$ chair conformation for the glucosamine residue (Figure 4.1). Further features of the anomeric carbon resonances of iduronic acid in O/N-desulphated heparin have come from studies of Fransson *et al.* (1978). Two forms of iduronic acid were observed (i.e. IdUA I and IdUA II) in O/N-desulphated heparin which showed the following features. The anomeric carbon of IdUA I had a chemical shift (99.75) and a (J_1(CH)) coupling constant (171 Hz) similar to those of sulphated IdUA (99.65; 173 Hz). It was resistant to periodate oxidation at pH 3.0 and 37°C when the adjacent glucosamine moieties were unsubstituted at C(2)-NH_2. These results together with pK_a data suggest that IdUA I has retained the $^1C_4$ conformation. The other IdUA form (II) had a chemical shift (102.35) similar to that of GlcUA (102.95). However, both the J_1(CH) coupling constant (165–170 Hz) and the apparent pK_a value (3.25) were significantly higher than those of GlcUA (163 Hz and 2.5 respectively). Since IdUA II was sensitive to periodate oxidation, the 2-OH and 3-OH groups should be equatorially disposed as in the $^4C_1$ and $^1S_3$ conformers.

Some characteristic features of the ^{13}C nmr spectrum of heparin-like polysaccharides are that they readily monitor the decrease in N-sulphation of the glucosamine residues and the O-sulphation of the uronic acid residues as downfield shifts in the respective C-2 and C-1 signals, as well as the changes in the methyl carbon peak at 23 ppm.

4.1.4 Periodate oxidation

Although periódate oxidation is commonly employed in the chemical structure analysis of sugar units it has received further attention recently in terms of its value in providing information on the conformation of sugar residues in polysaccharide chains. Scott and Tigwell (1978) have studied the kinetics of periodate oxidation of all the known glycosaminoglycan structures and have demonstrated marked differences in the second order rate constants or periodate oxidation.

It is well established that oxidative cleavage between C-2 and C-3 of the uronic acid residues in glycosaminoglycans is prevented or inhibited if the HO-2 and HO-3 1) have a diaxial configuration, 2) are sulphated, or 3) are involved in hydrogen bonding.

Heparan sulphate, which contains approximately 70% glucuronic acid, is rapidly oxidized. It appears that its glucuronic acid residues are being oxidized at a rate that approaches that of the monomer (Scott and Tigwell, 1978; Fransson, 1978). Fransson, in using this technique, confirmed that non-sulphated IdUA residues in heparin-related molecules adopt the $^4C_1$ conformation because the uronic acid residues were readily oxidized by periodate at pH 7.0 and 37°C. In contrast, periodate-resistant structures at pH 3.0 and 4°C were obtained with heparan sulphate and desulphated heparin. These results were interpreted in terms of a hydrogen bonding effect. With L-iduronic acid in the $^4C_1$ and $^1S_3$ conformers these residues were susceptible to periodate at pH 3.0 and 4°C provided adjacent glucosamine residues had free amino groups. On the other hand, L-iduronic acid residues in the $^1C_4$ chair form are resistant, together with D-glucuronic acid residues in combination with N-sulphated glucosamine residues as cooperative inter-residue hydrogen-bonding may be involved (Fransson, 1978).

4.1.5 Infra-red absorption and laser Raman spectroscopy

Infra-red absorption spectroscopy of solid phase samples has been used widely, based primarily on the analysis of weak bands in the 850–800 cm^{-1} region attributed to C–O–S vibrations (Spedding, 1964). From studies of O-sulphated monosaccharides and chondroitin sulphate, the bands at 850, 830 and 820 cm^{-1} have been assigned to axial, equatorial and primary (e.g. CH_2–O–S-structures) positions (Orr, 1954; Lloyd *et al.*, 1961ab; Bansil *et al.*, 1978). However, variations on the physical condition of the samples can produce broad absorption bands (10–20 cm^{-1}) and differences in peak position by as much as 20 cm^{-1}. Bettelheim (1970) has summarized the absorption bands of the glycosaminoglycan backbone.

The recent investigations of Cabassi *et al.* (1978) are the only reported studies on the infra-red absorption and laser-Raman scattering of the heparin-like polysaccharides. The infra-red spectra (Figure 4.5a) for aqueous solutions of heparin-like polysaccharides are dominated by the strong and complex absorption centred at ~1230 cm^{-1} and associated with the antisymmetric stretching vibrations of the S=O groups (ν_{as} S=O). The ν_{as} S=O band is strong and broad, and comprises at least two major components (Table 4.3). In the spectrum of heparin, the shoulder of the ν_{as} S=O band at 1180 cm^{-1} is attributable to the contribution of N-sulphate groups. This assignment was confirmed by the absence of an analogous shoulder in the spectrum of partially N-desulphated heparin.

A linear relationship has been found (Cabassi *et al.*, 1978) for the absorbance of the infra-red ν_{as} S=O band as a function of the content of sulphate groups for a number of heparins and a N-desulphated heparin, the sulphate content being determined by a conductimetric procedure (see Footnote 5.1).

Laser-Raman spectra of heparin have been obtained in aqueous solution and in the solid state (Figure 4.5b, Table 4.3). The most prominent Raman peak (at 1060 cm^{-1}) is attributable to the symmetrical vibration of the S=O groups (ν_s S=O), with N-sulphates emitting at somewhat lower frequencies (~1040 cm^{-1}) than O-sulphates. Raman spectra for solutions in water (and D_2O)

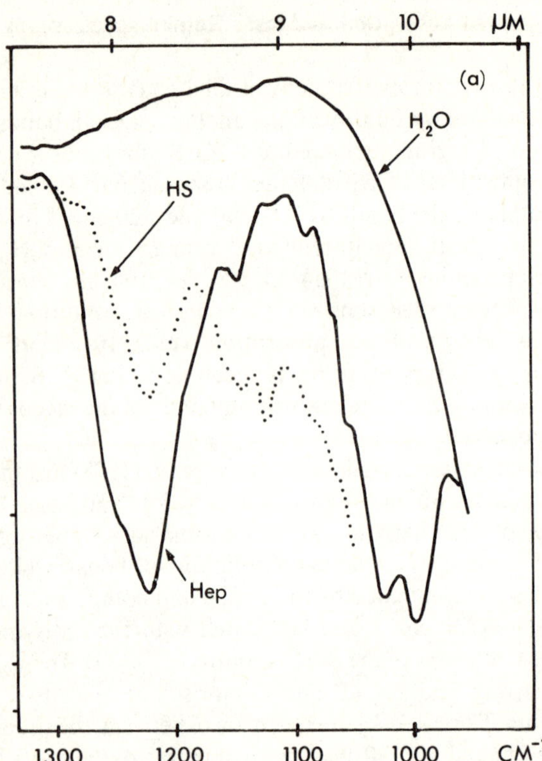

FIGURE 4.5 a) Infra-red spectra of commercial beef lung heparin in aqueous solution, and a heparan sulphate sample from unknown source. (Adapted from Cabassi *et al.*, 1978; reproduced with permission of the publishers.)

show a number of medium intensity peaks, two of which are usually prominent. However, no apparent relationship was found between these maxima and the type or orientation of sulphate groups, which stresses the need for caution in using laser-Raman frequencies in this region for correlation purposes.

Bansil *et al.* (1978) have measured the infra-red spectra of films cast from the same samples for various glycosaminoglycans (non heparin-like) as used in Raman spectra determinations. They identified the sensitivity of the Raman spectra to the orientation of the sulphate group relative to the pyranose ring together with

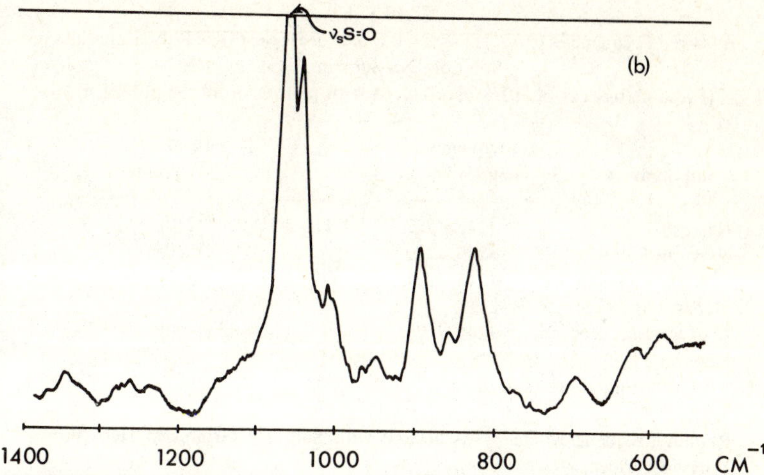

FIGURE 4.5 b) Raman spectra of commercial beef lung heparin in aqueous solution. (Adapted from Cabassi *et al.*, 1978; reproduced with permission of the publishers.)

C(1)–H deformation mode of α and β anomers occurring in the range 884–899 cm^{-1} for chondroitin 6-sulphate, chondroitin 4-sulphate and hyaluronate where β-linkages occur. Although little data have been published on heparin-like polysaccharides utilizing laser Raman spectra, these studies do focus on the potential of this experimental procedure in analyzing conformational states of sugar residues and their substituent in association with simple and poly-counterions.

4.2 CHAIN CONFORMATION

4.2.1 Viscosity

It appears that only a limited amount of information can be derived from hydrodynamic measurements on polyanions as the theoretical interpretation of their behaviour is complex. Several interpretations directed at the hydrodynamic properties of heparin have demonstrated that it probably exists in solution as an essentially random flexible coil. Its flexibility is governed

TABLE 4.3

Infra-red Frequencies (cm^{-1}) of ν_{as} S=0 Bands of Heparin-like Polysaccharides in Solution (Na Salts in H_2O)

(From Cabassi *et al.*, 1978; reproduced with permission of the publishers.)

Compound	Frequency (cm^{-1})		Band Barycenter
Heparin	1245 *1222*	~1115(sh)	(1230)
Heparan Sulphate	1245 *1220*		(1220)
N-Desulphated Heparin	1245 *1225*		(1235)
COOH-reduced Heparin	1245 *1220*		(1230)

* The strongest peaks in italic.

Raman Frequencies (cm^{-1}) Associated with Sulphate Groups in Heparin-like Polysaccharides

Compound	ν_S S=0 (cm^{-1})	ν_{as} C–O–S[a] (cm^{-1})	
Solution			
Heparin	1062	893	827
Solid phase			
Heparin	1065	[b]	[b]
Heparan sulphate	~1055	[b]	[b]

[a] Tentative assignment. The vibration is thought to be strongly coupled with other vibrational modes.

[b] Data not obtained.

by micro-ion interaction along its length and by the conditions of the solvent, such as ionic strength and pH; this behaviour is typical of most polyanionic electrolytes studied so far.

A simple interpretation of conformation type has been utilized through the Mark–Houwink relationship where the intrinsic viscosity $[\eta]$ is correlated to the molecular weight (M) by the relationship

$$[\eta] = KM^a \qquad 4.1$$

TABLE 4.4
Constants of the Mark–Houwink Relation for Heparin
($[\eta]=K M^a$)

Solvent	$K \times 10^3$ (ml/g)	a	Reference
0.1M NaCl pH 6.5	1.75	0.98	Lasker and Stivala (1966)
0.5M NaCl pH 2.5	7.85	0.80	Lasker and Stivala (1966)
H_2O	0.131	1.59	Liberti and Stivala (1967)
0.01M NaCl	8.06	1.13	Liberti and Stivala (1966)
0.1M NaCl	1.54	1.00	Liberti and Stivala (1966)
0.5M NaCl	3.55	0.9	Liberti and Stivala (1966)
0.2M NaCl	1.58	1.00	Laurent (1961)

Here K is a constant, and the exponent a is a constant characteristic of the shape of the molecule. The values for K and a that have been measured on heparin samples are given in Table 4.4. The fractions used in these analyses were not fractionated on the basis of molecular weight and therefore are unlikely to be monodisperse. No relationship in the form of the Mark-Houwink equation has been published for heparan sulphate preparations.

The differences in values of the exponents reflect changes in solute–solvent interactions due to variation of pH and ionic strength. For free-draining coils, the value of the exponent in the Mark–Houwink equation is 0.8 to 1.0, while a value of 0.5 corresponds to a compact solvent-immobilizing coil and a value of 1.8–2.0 for rod-like chains. The relatively high values of the exponents at high ionic strength (>0.1) suggest that heparin approximates to the free-draining random coil model. Based on the values of the Mark–Houwink exponent it would appear that heparin in water is asymmetrical, approaching spherical symmetry as the ionic strength of the system increases.

Studies of the variation of $[\eta]$ with pH have shown varying trends in the pH range in which the carboxyl groups are ionising. In general, as they ionise at pH values greater than 2.0m the $[\eta]$ increases, since in this range the sulphate groups should be fully ionized. The variation noted for intrinsic viscosity as a function of pH is most likely due to the ionic strength at which these changes take place (Section 5.1.4). Stivala and Liberti (1967) found maximal changes in $[\eta]$ in the pH range of 2.0 to 3.3 at 0. 1M NaCl, whereas Chung and Ellerton (1976) found a significant change of $[\eta]$ over the pH range of 2.44 to 5.28 in 0.01 m NaCl.

Measurement of both intrinsic viscosities and sedimentation coefficients permits the estimation of molecular size and shape parameters in solution, by the use of the Scheraga and Mandelkern equation,

$$B = \frac{NS[\eta]^{1/3}\eta_0}{M^{2/3}(1-\bar{v}\rho)} \qquad 4.2$$

where B is a constant dependent on axial ratios for ellipsoids of revolution, N is Avogadro's number, S the sedimentation coefficient, $[\eta]$ the intrinsic viscosity, η_0 the solvent viscosity, M the molecular weight, $\bar{v}$ the partial specific volume and ρ the solution density. Values of B for heparin fractions, measured in 0.5 M NaCl, pH 2.0, are constant and agree reasonably well, within experimental error, with the theoretical value of 2.5×10^6 for a random coil model (Lasker and Stivala, 1966; Lages and Stivala, 1973; Stivala and Ehrlich, 1974). There is, however, a significant lack of information on hydrodynamic measurements of heparin at physiological conditions.

Viscosity measurements of heparin at low ionic strength or in water have been treated using the relation

$$\eta_{sp/c} = [\eta]_\infty(1 + k'/\sqrt{c}) \qquad 4.3$$

in an effort to separate out the factors controlling the overall conformational changes. The shielded intrinsic viscosity $[\eta]_\infty$ is obtained by extrapolation of the data to infinite concentration and the electrostatic interaction parameter k' is obtained from the slope. The values of $[\eta]_\infty$ correspond well to the values of the intrinsic viscosity of uncharged parent polymers and good agreement is found on comparison of $[\eta]_\infty$ with $[\eta]$ values using 0.1 M NaCl as salt (Yuan and Stivala, 1972). Assumption of random coil behaviour for the shielded polyion, and application of the Flory–Fox relationship

$$[\eta]_{\infty} = 2.1 \times 10^{21} \, (\bar{r}^2/M)^{3/2} \qquad 4.4$$

to obtain the root-mean-square end-to-end distance $(\bar{r}^2)^{1/2}$ for sample $M_w = 12{,}300$ give $(\bar{r}^2)^{1/2} = 10.5$ nm. For a random coil, the radius of gyration (R_g) is given by $R_g^2 = (1/6)\bar{r}^2$ which yields $R_g = 4.28$ nm. This result agrees adequately with the value of 3.59 nm obtained from low angle X-ray scattering from moderately concentrated solutions (0.25–1.00% heparin in water) (Stivala *et al.*, 1968).

For a polymer with a large number of statistical units k' of eqn. 4.3 is defined as

$$k' \simeq 5.22\sqrt{P/A^2} \qquad 4.5$$

where $P = M_1/h_1$, i.e. the molecular weight of a basic unit of length h_1, and A is the equivalent freely-jointed segment length when the backbone of the polyion is represented by a Gaussian coil of N segments. In this case k' depends only on the properties of the polymer backbone and not on the electrical properties of the polyion or the solvent. Taking a value of $A = 2.12$ nm for heparin (the value of the persistence length obtained from low angle X-ray scattering) calculated values of k' from equation 4.5 were in agreement with the observed k' values determined by equation 4.3 for heparins in the molecular weight range 6×10^3–13×10^3 (Yuan *et al.*, 1972).

In principal, the value of k' is a function of intramolecular interactions of the molecule, particularly hydrogen-binding. However, interpretation of this parameter is complex when changes in intramolecular configuration are invoked. Yuan and Stivala (1972) have found that values of k' were independent of dielectric constant (D) in the range $54.7 < D < 93.2$, whereas $[\eta]_{\infty}$ of heparin increases linearly with increasing dielectric constant of the solvent. Thus, for urea at 6M ($D = 93.2$) it is suggested that destruction of any existing hydrogen bonds may cause the chain to expand without change in k' while, on the other hand, the observation that both $[\eta]_{\infty}$ and k' decrease with increasing desulphation has been suggested to be due to the formation of intramolecular binding through either the hydrogen bonding of the unmasked amino groups or electrostatic interaction involving a protonated amino group (Stivala, 1977). A further complication is that, when the urea concentration exceeds 6M, the viscosity of heparin decreases with time and the time dependence is removed by adding salt (Yuan, 1974).

Some questions remain as to the value of quantitative viscometric analysis of heparin in water by capillary viscometers, as Chung and Ellerton (1976) have recently shown that heparin solutions unexpectedly exhibit shear dependence. The viscosity was found to decrease with increasing shear stress. The shear dependence was greatest in the absence of salts and became relatively independent at ionic strengths >0.01 in the shear stress range between 0.0193 and 0.222 dyne cm^{-2}. This suggests that even at very low shear stresses, some alignment of the molecules with the direction of the streamlines of flow has occurred, this alignment becoming less as the molecules become more flexible when the ionic strength is increased.

4.2.2 Diffusion and sedimentation

As in viscosity measurements, means for assessing chain flexibility from the sedimentation coefficient (S_0) can be performed by evaluation of constants (K_s) and exponents (γ) from the relationship $S_0 = K_s M^{\gamma}$. Lasker and Stivala (1966) have performed sedimentation velocity studies on bovine heparin fractions separated on an ECTEOLA column, and obtained values of $K_s = 1.2 \times 10^{-1}$ and $\gamma = 0.44$ with solvent 0.5M NaCl, pH 2.5. Theory predicts that for a compact coil $S_0 \propto M^{0.33}$ while for a free draining coil $S_0 \propto M^0$. For flexible polymers S_0 should vary as $M^{0.5}$ in a solvent and roughly as $M^{0.45}$ in a better solvent.

Diffusion and sedimentation measurements of heparin-like polysaccharides by analytical ultracentrifugation are shown in Table 4.5. Jamieson and Maret (1973) cite previous investigations, in which they have used quasi-elastic laser light scattering and obtained an empirical relation for the diffusion coefficient $D = 1.33 \times 10^{-4} M^{-0.59}$ from the plot of log D versus log M for three glycosaminoglycans, namely heparin, dermatan sulphate and

TABLE 4.5
Diffusion and Sedimentation of Heparin-like Polysaccharides

Sample	S_0 (Svedberg)	$D_0 \times 10^7$ ($cm^2\ sec^{-1}$)	Molecular weight	Solvent conditions	Reference
Heparin	3.06	9.8	15,800	1M KCl	Patat and Elias (1959)
	3.20	11.6	13,800	1M KCl	Patat and Elias (1959)
Heparin	—	7.95	11,900	N.I.	Barlow *et al.* (1961)
Aortic Heparan Sulphate	2.28	2.7	50,000	0.15M NaCl	Öbrink *et al.* (1975)
	2.70	3.0	50,000	0.05M $CaCl_2$	Öbrink *et al.* (1975)
Mactin A (*Spissula* (*Mactra*) *solidissima*)	2.65	4.41	24,800	N.I.	Burson *et al.* (1956)
Mactin B (*Cyprina* (*Arctica*) *islandica*)	2.57	3.70	28,700	N.I.	Burson *et al.* (1956)

Abbreviations: N.I. = Not identified.

chondroitin sulphate. Heparin of molecular weight $\simeq$ 11,000 has a $D \simeq 5.7 \times 10^7$ cm^2 sec^{-1}. Solvent conditions for these measurements were not published, and no other detailed information has been forthcoming.

4.2.3 Optical rotation

The optical rotatory power of glycosaminoglycans at 589 nm has been used extensively in their characterization and identification (Section 2.3.1) although its potential as a tool to investigate heparin-like polysaccharides has been handicapped in terms of what was a previous lack of knowledge of the chemistry of these compounds.

A particular point of interest, in this context, which has received little attention in terms of the heparin-like polysaccharides is the effect of polysaccharide conformation on optical rotation. Some circumstantial evidence for $[\alpha]_D$ dependence on heparin conformation has come from the studies of Hirano (1971) where he found that the optical rotatory power of heparin was markedly reduced in the presence of urea ($[\alpha]_D^{21}$ (H_2O)$=45°$; $[\alpha]_D^{21}$ (7M Urea)$=9°$) and this was interpreted in terms of heparin existing as a hydrogen-bonded structure in water.

4.2.4 Ultraviolet circular dichroism (CD)

Early studies by Stone (reviewed in 1969, 1977) had demonstrated that heparin and other glycosaminoglycans show intrinsic Cotton effects in the ultraviolet region. All glycosaminoglycans, with the possible exception of dermatan sulphate (Park and Chakrabarti, 1978), show a negative CD band around 210–212 nm (Figure 4.6). The assignment of this band has been tentatively ascribed to n-π* transitions of the amido groups of the hexosamine residue for the following reasons (Stone, 1969); 1) relatively low values for the CD bands on glucuronate, 2) the CD behaviour of N-acetyl glucosamine, 3) the variation of relative differences in CD among the salt and acid forms and 4) the fact that keratan sulphate, while showing a similar CD spectrum to heparin, does not have a uronate moiety. More recent studies by Park and Chakrabarti (1977, 1978) on the effects of varying pH on the circular dichroism of heparin and

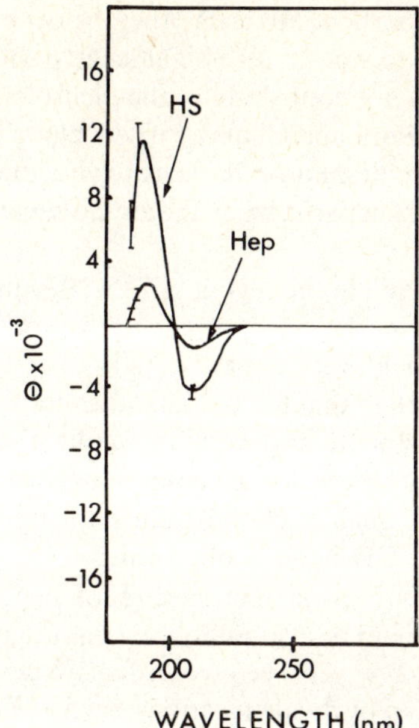

FIGURE 4.6 Ultraviolet circular dichroism of heparin and heparan sulphate where θ denotes the molar ellipticity (in degree–cm^2/decimol) based on the average formula weight per hexosamine residue. (From Stone, 1971; reproduced with permission of the publishers.)

other glycosaminoglycans have revealed that CD changes in the wavelength region of 210–230 nm with different pH values can be qualitatively related to pK_a values of the carboxyl groups in these polymers. It is, therefore, apparent that the carboxyl chromophore contributes significantly to the dichroic behaviour of these polymers in this wavelength region. Park and Chakrabarti have found that the iduronic acid residue carboxyl chromophore dichroic peak is located near 210 nm, whereas the D-glucuronic acid residue has a small ellipticity at this wavelength. The apparent value of the pK_a of iduronic acid residue determined at 210 nm is in agreement with values obtained by other techniques (Table 5.2). At 230 nm, both the iduronic acid residues and the glucuronic acid residues show changes in dichroic behaviour with pH; for pH values above 4.5 the ellipticity changes of heparin at both 210 nm and 230 nm are proportional to each other; below pH 3.5, the CD spectra of heparin do not change significantly so that sulphate groups probably do not contribute to the dichroic spectra in this wavelength region. Park and Chakrabart concluded that the n-π^* transition of the amide and/or the carboxyl group is relatively small, as in the case of heparin, when the amino sugar is linked with iduronic acid.

The second CD band in the region of 185–192 nm in Figure 4.6 has been assigned to the π–π^* amide transition (Stone 1965, 1967). Its sign and magnitude appear to reflect two distinct classes of glycosaminoglycan in that for 1–4 linked amino sugars, such as heparin, heparan sulphate and keratan sulphate, a positive CD band is obtained, whereas for glycosaminoglycans having 1–3 linked amino sugars, such as chondroitin sulphate and dermatan sulphate, a negative CD band is obtained.

No significant changes in CD spectra of heparin-like polysaccharides, which could be attributable to conformational changes in the molecule, have yet been recorded. When the sodium counteriön is replaced by divalent cations, such as Ca^{2+} and Ba^{2+} little change in CD spectra is seen (Öbrink *et al.*, 1975; Chung and Ellerton, 1976). Marked changes in the CD spectra of heparin have been recorded for Cu^{2+} complexation under certain conditions (Chung and Ellerton, 1976; Mukherjee *et al.*, 1978) although it is not clear what these changes represent.

4.2.5 Induced Cotton effects with methylene blue

It has been known for many years that complexes of heparin and methylene blue at polyanion charge/dye molar ratios of unity, give rise to induced Cotton effects characterized by a symmetric double CD spectrum (Stone, 1965, 1969; Stone and Moss, 1967). In view of the original findings of Blout and Stryer (1959) that helical forms of poly-amino acids and nucleic acids displayed characteristic intrinsic Cotton effects while the nonhelical or random coil forms did not, Stone concluded that the effect with heparin-like polysaccharides was attributed to the binding of the metachromatic dye to a left-handed helical array of negatively charged sites on the polysaccharide molecule. More recently, the CD spectrum has been shown to fit that theoretically predicted for the exciton-like model for dipole coupling of dyes in a helical array (Stone, 1977). The nature of the binding of the dye to the polysaccharide depends on dye–polyanion interaction and asymmetric (with respect to polyanion) dye–dye interaction. It is known that disruption of the array by redistribution of dyes among excess sites (Stone, 1977), by dissociation of dyes due to competition with $MgCl_2$, or by limited hydrolysis of the anionic sites (N-desulphation) led to a diminution of the induced Cotton effects.

Stone (1969) has found similar patterns of intrinsic Cotton effects with heparin, heparan sulphate and keratan sulphate, whereas chondroitin 4-sulphate, chondroitin 6-sulphate and dermatan sulphate show only a single positive CD band. Although the latter group of glycosaminoglycans has been shown capable of forming helices in solid state, as demonstrated by X-ray diffraction studies (see Section 4.1.1), interpretation of their behaviour characterized by their CD spectra has yet to be forthcoming. In any case, for heparin-like polysaccharides the possibility that ordered regions may be induced into the molecule, not necessarily extended through the whole chain, appears likely. This will depend on environmental conditions such as in condensed phases or through the co-operative interactions of bound counterions.

4.2.6 Electron microscopy

The only work published on electron microscopical studies of heparin molecules is by Hirano (1972) who found that the heparin molecules were linear and 16 ± 0.4 nm in length. The source of heparin was a commercial hog mucosal preparation.

4.2.7 Polymolecularity of single chains

It is well recognised that commercial heparin preparations show a wide spectrum of molecular weights; often reported to be in the range of 6,000 to 30,000 (Table 2.1). It is of interest to determine whether the distribution of molecular weights is a continuous one (polydisperse) or discontinuous (paucidisperse) in the light of the requirements of the biosynthetic process which includes chain termination and proteoglycan degradation as catalyzed by an endoglycosidase (Section 4.3). At the present time, these processes are not well understood in terms of heparin or heparan sulphate polymolecularity.

Further problems are also encountered with molecular weight determinations of heparin-like polysaccharides due to the marked non-ideality exhibited by these compounds as generated by their strong polyelectrolyte nature. Non-ideality effects can be overcome to some extent by the use of high concentrations of ambient simple electrolyte and by increasing the valence of the counterion (Öbrink *et al.*, 1975). There is no evidence forthcoming to suggest that non-ideality may be, in part, due to chain entanglement as proposed by Jaques and McDuffie (1978).

Gel filtration chromatography has been shown to be an effective method for demonstrating size heterogeneity of heparin, there being a logarithmic relationship between molecular weight and elution volume (Johnson and Mulloy, 1976) (see also Table 7.3). (Often under the heavy loading conditions used with the preparative columns, the elution patterns do not match those obtained with small loads on an analytical column (Johnson and Mulloy, 1976; Laurent *et al.*, 1978).) Equilibrium sedimentation analysis has only been used with limited success in estimating the polymolecularity of heparin preparations (Lasker, 1977). Analysis of diffusion curves in terms of Gaussian behaviour suggested that preparations were not detectably heterogeneous (Lasker, 1977). In an earlier study by equilibrium methods, Braswell (1968) found an average ratio of Z-average to weight-average molecular weight of 1.8. On the basis of theoretical analysis of polymerization by linear condensation or by random cleavage of a uniform highly polymerized molecule the ratio is predicted to be only 1.5. He concluded that there are several unresolvable but distinctly different species present.

Although equilibrium sedimentation is probably the best method available for determining mean molecular weights of heparin, there is a useful body of literature which correlates measurements of this kind with limiting viscosity. The equation relating intrinsic viscosity $[\eta]$ to molecular weight may be expressed in the form of equation 4.1, and the exponent a may vary between 0.8 and 1.0 depending on ionic strength and pH (see

Section 4.2.1). Another technique which is commonly used for molecular weight determination, but also requires precalibration, is polyacrylamide gel electrophoresis (Hilborn and Anastassiadis, 1971).

Previous reports (Nader *et al.*, 1974; McDuffie *et al.*, 1975) have purported that heparin can be fractionated by isoelectric focusing into at least 21 components, with 'pI values' in the pH range of 3.2–4.2. This fractionation was suggested to be associated with the differences in molecular weight, the low pI components exhibiting the smallest size ($M = 3{,}000$) while the high pI components had the highest molecular weight ($M = 37{,}500$), with graded intervals of approximately 500–2,000 in size. Although these reports of isoelectric focusing of heparin into 21 bands stressed a causal relationship between a 'fractionation' of highly polydisperse molecular weight components and banding, subsequent studies (Johnson and Mulloy, 1976; Righetti and Gianazza, 1978) showed that an effect of the molecular size of heparin was seen only in a non-specific fashion between the molecular sizes at the extreme ends of the distribution. In general, the isoelectric focusing of purified molecular weight fractions by gel chromatography of heparin (Johnson and Mulloy, 1976) was not specifically related to molecular weight. The basis for the anomalous fractionation has now been shown to be due to an interaction between the polysaccharide and different amphoteric species in the Ampholine mixture (Righetti and Gianazza, 1978; Righetti *et al.*, 1978). Ampholine is a mixture of carrier ampholytes which are the buffering amphoteric compounds used to generate and stabilize the pH gradient in isoelectric focusing.

The polymolecularity of heparin preparations can be described by a plot of molecular weight versus weight fraction for fractionated material. In this manner the integral dispersion curves for heparin in Figure 4.7 have been constructed from molecular weight data (as determined by equilibrium sedimentation analysis) of fractionated heparin, published by Laurent *et al.* (1978) (Table 7.3). The two distinct peaks present in the differential curve for heparin are suggestive of two discrete species in the unfractionated preparation with the possibility of a third species in the middle region. The high molecular weight fraction ($M_w = 25{,}000$, ($\Delta\%$

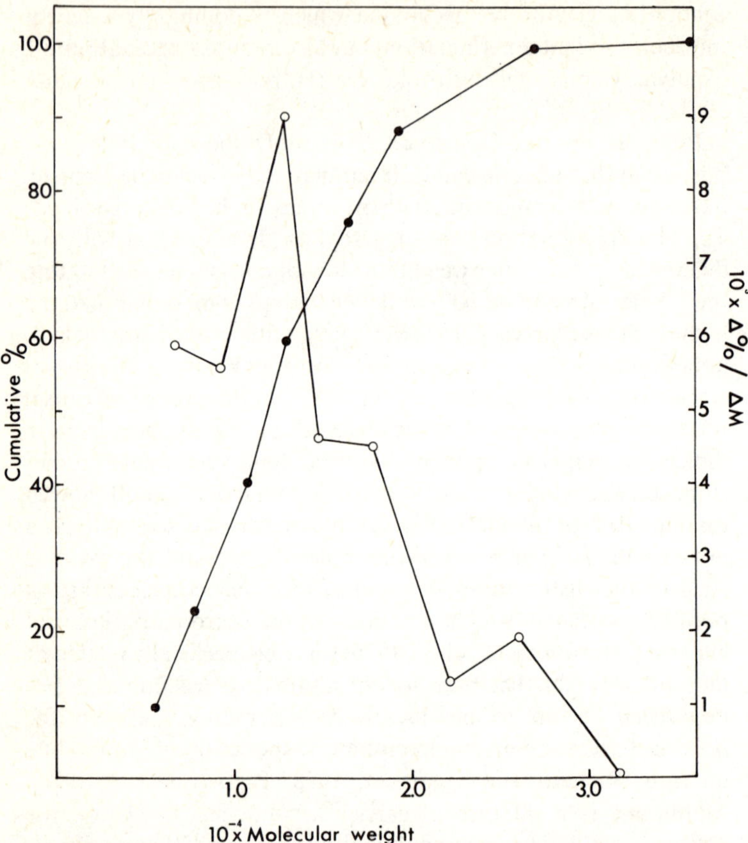

FIGURE 4.7 The integral dispersion curve of commercial heparin, fractionated by analytical gel chromatography with molecular weights determined by sedimentation equilibrium, calculated from data of Laurent *et al.*, 1978 (Table 7.3). Plots are made of cumulative percent versus molecular weight (●——●) and relative change of cumulative percent relative to molecular weight ($\Delta\%/\Delta M$) versus molecular weight (○——○).

$/\Delta M_w)_{max}$) probably corresponds to products with peptide core present with one or more chains or oligosaccharides bound (Section 4.3). Similar analysis by Lasker (1977) was performed on a commercial heparin preparation which was alcohol-fractionated and the intrinsic viscosity of the fractions measured.

Notwithstanding the low average molecular weight value of the preparation ($M_w = 8{,}940$), integral dispersion analysis demonstrated the existence of a bimodal distribution of material of molecular weights near 12,000 and 10,500.

4.3 MULTI-CHAIN FORMS

4.3.1 Heparin

Although there is much less information concerning the multi-chain form of heparin (often referred to as macromolecular heparin or heparin proteoglycan) the evidence now suggests that heparin is synthesized as a complex of very high molecular weight, most probably consisting of a protein core with pendant heparin chains covalently attached. This structure is then degraded in some tissues, intracellularly, into a heterogeneous population of short polysaccharide chains by means of an endoglycosidase. These degradation products are primarily single polysaccharide chains with a molecular weight ranging from 5,000 to 15,000 and representative of the common commercial heparin preparations.

Early studies suggested that some of these polysaccharides contain residual peptides when isolated after proteolytic digestion of tissues (Lindahl *et al.*, 1965; Lindahl, 1966). The peptides are linked to the polysaccharides proper by a neutral trisaccharide – (D-galactosyl)–(D-galactosyl)–(D-xylose) with the hydroxyl group of serine (Table 2.1). This type of linkage is characteristic of the majority of linkage regions of the various glycosaminoglycans (with the exception of hyaluronate and keratan sulphate) with protein cores (for review see Lindahl and Rodén, 1972). These conclusions are primarily based on the susceptibility of the linkage to alkali, which causes the liberation of single polysaccharide chains by a β-elimination mechanism (Anderson *et al.*, 1965) with concomitant destruction of the serine residue to form β-alanine after reduction. It appears that not all the serine residues are alkali-sensitive as shown in studies on fractions obtained from mouse mastocytoma (Ögren and Lindahl, 1971; Chandrasekran *et al.*, 1975; Robinson *et al.*, 1978). There is also conjectural evidence that threonine may participate in protein core polysaccharide chain linkage in heparin preparations from bovine liver capsule (Serafini-Fracassini *et al.*, 1973).

Lloyd *et al.* (1967) first demonstrated the high molecular weight form of heparin. They used extracts of mast cell granules from rat peritoneal cells in which the heparin was excluded from Sephadex G-100. On the other hand, commercial heparin easily penetrated the gel. After treatment with alkali or proteolytic enzymes, the heparin material behaved on gel chromatography as commercial heparin samples. The macromolecular nature of heparin was established by Horner in 1971 who isolated a preparation from pronase digested rat skin with a molecular weight of 1.1×10^6 (see also Robinson *et al.*, 1978). This macromolecular heparin had a sedimentation coefficient of 12.8×10^{-13} sec^{-1}, an intrinsic viscosity of 270 ml g^{-1} (as compared to a value of 17 ml g^{-1} for pig mucosal heparin) and an anticoagulant activity of 123 IU/mg.

High molecular weight heparins have been obtained from several tissue sources of Cynomolgus monkeys including lung, liver and skin (Horner, 1976). A heparin proteoglycan with a molecular weight of 750,000, has been obtained from rat peritoneal mast cells by non-proteolytic procedures (Yurt *et al.*, 1977).

The heparin chain size of these high molecular weight heparins has been estimated subsequent to alkali digestion. Molecular weights for these chain preparations vary from 40,000, as estimated by gel chromatography of alkali treated rat peritoneal mast cell preparations (Yurt *et al.*, 1977) to 60,000 to 100,000 for rat skin preparations (Robinson *et al.*, 1978) as estimated by ultracentrifugation analysis, by the amount of serine destroyed on alkali treatment or by alkali treatment in the presence of sodium [^{3}H]-borohydride and measurement of [^{3}H]-xylitol formed. Horner (1971) has obtained lower molecular weights, of the order of 36,000, for rat skin preparations as estimated by end group analysis and serine content. In any case, these studies clearly demonstrate that the individual polysaccharide chains in macromolecular heparins are considerably longer than the single chain heparin molecules that occur in certain tissues and in commercially available preparations.

The nature of the central core of these high molecular weight heparins has been a debatable issue. Horner (1971) suggested, on the basis of the degradation pattern of macromolecular heparin by ascorbic acid, that macromolecular heparin consists of individual polysaccharide chains held together by a core structure that is possibly of polysaccharide nature (resistant to ascorbate and pronase degradation). Subsequently, all evidence has pointed towards a protein core for high molecular weight heparins, although the original results of Horner (1971) remained to be fully explained.

Horner's (1971) initial preparation of macromolecular heparin from pronase digested rat skin exhibited a protein content of 0.35%, and contained only serine and glycine residues in almost equal molar proportions. In a more recent study (Robinson *et al.*, 1978) on the nature of the core material, rat skin macromolecular heparin was degraded with nitrous acid to give a core fraction of molecular weight about 20,000 as determined by sedimentation equilibrium. Chemical analysis of this core fraction demonstrated that it contained more than 90% of the neutral sugars (xylose and galactose) and amino acids (essentially serine and glycine) of the total reaction products but less than 4% of the uronic acids. Analytical data for the core fraction indicated about 15 carbohydrate-peptide linkage regions/molecule and, in addition, 1 or 2 uronosyl-N-acetyl-glucosamine disaccharide units for each linkage region. It was proposed that rat skin heparin contained a poly(–ser–gly–) polypeptide core in which 2 out of 3 serine residues are substituted with polysaccharide chains accounting for at least 10 polysaccharide chains, each having a molecular weight of 60,000 to 100,000. In accord with this analysis, Figure 4.8 shows a tentative structure of the heparin proteoglycan. The fact that this type of molecule is resistant to several proteolytic enzymes including pronase (Yurt *et al.*, 1977; Horner, 1971) does suggest some steric protection of the core by densely packed pendant heparin polysaccharide chains.

Amino acid analysis of various commercial preparations of heparin has demonstrated a large variation in the amino acid content (Wang and Jaques, 1974) with no particular correlation with anticoagulant activity. The lack of consistency in the amino acid compositions of these preparations does suggest that impurities may copurify with the commercial preparations of

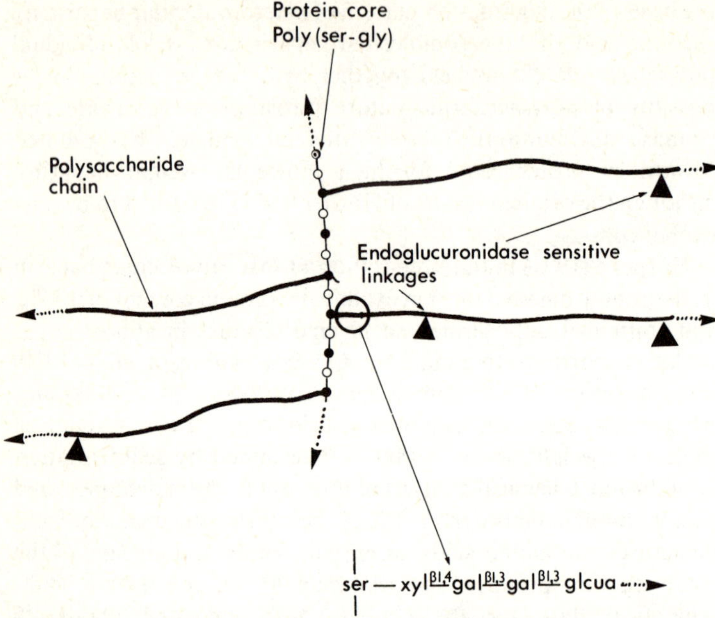

FIGURE 4.8 Tentative structure of a portion of the heparin proteoglycan from rat skin. (See text for details.)

heparin. Preparations can be obtained which lack serine. Normal preparations have a peptide or protein content in the range of 0.2–3%. It is probable that extraneous protein may be present in these preparations which may affect their properties.

In earlier studies, Serafini *et al.* (1973) proposed that the bovine liver capsule heparin contained on e residue of xylose/polysaccharide chain and that most of the polysaccharide preparation occurred as doublet molecules, consisting of two heparin chains linked to a common polypeptide. In contrast, Jansson *et al.* (1975) in studying 80% of the total heparin isolated by 4M GuHCl demonstrated that the heparin occurs as single polysaccharide chains. Further, most of these molecules were shown to lack the polysaccharide–protein linkage region. They had a reducible, terminal uronic acid residue. It was proposed that a major portion of the liver capsule heparin is a degradation product released from a large molecule by an endoglycuronidase. Only about 15% of these chains contain the polysaccharide–protein linkage region. The products of this endoglycosidase action will yield fragments similar to commercial heparin.

This has been confirmed in the work of Robinson *et al.* (1978) in which heparin chains obtained from alkali-treated macromolecular rat skin heparin were incubated with a mastocytomal endoglycosidase to yield partially depolymerized products with a molecular weight approaching that of commercially available heparin (7,000 to 25,000). Actual degradation of the multi-chain form of the proteoglycan was not studied. Horner (1977) has demonstrated that after the injection of [^{35}S]-sulphate into mice or monkeys, the label gave identical specific activity for high and low molecular weight forms of heparin. This further demonstrates the possible existence of a product–precursor relationship between these two forms. Heparin depolymerizing endoglycosidase has been shown to be present in murine Firth mastocytoma tissue (Ögren and Lindahl, 1975), in rat liver lysosomal preparations (Arbogast *et al.*, 1977) and in rat spleen (Höök *et al.*, 1977); the enzymes having optimal activity at pH 4.4.

In contrast to these patterns of degradation as exercised by the endoglycosidase in mouse mast cell tumours, Yurt *et al.* (1977) have found in studies of the time-related growth in size of [^{35}S]-heparin from rat peritoneal mast cells, that the proteoglycan appears to become progressively larger with time of incubation from 1–4 hours.

In summary, it appears that an endoglycosidase is capable of degrading the individual polysaccharide chains of macromolecular heparin, by cleaving specific linkages located in the peripheral portions of the molecule relative to the polysaccharide–protein linkage region. As a result, some of these chains have a full complement of polysaccharide–protein linkage region components (serine or serine-containing peptides, xylose and galactose), whereas others are deficient in this regard and carry a glucuronic acid residue at their reducing terminals. Several problems remain concerning the activity of the endoglycosidases in various tissues, the localization of this degrading activity and the nature of degradation of the protein core *in vivo* and possibly during proteoglycan isolation.

4.3.2 Heparan sulphate

Early investigations had established the proteoglycan nature of heparan sulphate from human aorta (Knecht *et al.*, 1967). Table 4.6 summarizes the results of recent investigations on the isolation and properties of heparan sulphate proteoglycans from various tissues and cell plasma membranes. The reported molecular weights and corresponding chain sizes of these preparations appear to vary widely and may depend on the method of extraction (Table 4.6).

The localization of these macromolecular heparan sulphates appears to be confined primarily to cell-surface-associated-material and material existing in extracellular environments. High molecular weights forms of heparan sulphate have also been isolated from microsomal preparations (Oldberg *et al.*, 1979). Further, it is noteworthy that portions of the cell surface-associated heparan sulphate are salt extractable (Oldberg *et al.*, 1977, 1979). Several investigations have suggested that extracellular heparan sulphate is intimately associated with elastin in the extracellular matrix. These conclusions have been based on the selective solubilization of heparan sulphate by elastase treatment of the parenchyma and pleura of bovine lung (Wusteman, 1972) and bovine aorta (Radhakrishnamurthy *et al.*, 1977). However, the possibility of non-specific protease activity in these treatments has not been completely eliminated.

The amino acid compositions of heparan sulphate proteoglycan preparations derived from several sources are shown in Table 4.7. It is evident that serine, glycine, glutamine (glutamic acid) and asparagine (aspartic acid) are the major and common amino acids of these preparations, whereas cysteine residues are lacking. Preparations from bovine aorta and human dentine also have high contents of alanine. On the basis of alkaline degradation studies, it has been suggested that serine residues probably account for the major carbohydrate linkage to the protein core (Radhakrishnamurthy *et al.*, 1977; Oldberg *et al.*, 1979). A number of unusual linkages may be involved through the participation of threonine residues (Oldberg *et al.*, 1979) and testicular hyaluronidase-susceptible linkage regions (Kraemer and Smith, 1974). Molecular models for heparan sulphate proteo-

TABLE 4.6
Sources and Some Physical Properties of Heparan Sulphate Proteoglycans

Source	Extraction Method	Molecular Weight	Buoyant Density in CsCl + 4M GuHCl (g/ml)	Chain Size from alkaline treatment	Reference
Ovine brain	Dissociative extraction with 4M GuHCl	25,000–30,000	—	—	Branford-White and Hardy (1977)
Bovine aorta	Elastase solubilized	64,000	1.74	—	Radhakrishnamurthy *et al.* (1977)
Chinese hamster cells	Trypsin digestion	175,000	—	44,000	Kraemer and Smith (1974)
Swiss mouse 3T3 cells	Trypsin digestion	750,000	—	—	Johnston *et al.* (1979)
Ascites Hepatoma AH66 Plasma Membranes	Detergent solubilization	60,000	—	—	Mutoh *et al.* (1978)
Rat liver membranes	Detergent solubilization	75,000	> 1.55	14,000	Oldberg *et al.* (1979)

TABLE 4.7
Amino Acid Composition of Heparan Sulphate Proteoglycans
(Residues per 1,000 residues)

Amino Acid	Bovine Aorta[1]	Human Dentine[2]	Rat Liver Microsomes[3]
Cysteic Acid	—	—	—
Aspartic/NH_2	115	109	125
Hydroxyproline	—	—	—
Threonine	32	96	78
Serine	96	158	142
Glutamic/NH_2	117	124	161
Proline	49	21	77
Glycine	267	169	100
Alanine	235	114	57
Valine	133	13	33
Cysteine	—	—	0
Methionine	0	22	15
Isoleucine	90	52	34
Leucine	74	26	68
Tyrosine	11	10	19
Phenylalanine	29	7	16
Lysine	22	52	41
Histidine	11	12	12
Arginine	8	15	21

References
[1] Radhakrishnamurthy *et al.* (1977)
[2] Branford-White (1978)
[3] Oldberg *et al.* (1979)

glycans from bovine aorta (Radhakrishnamurthy *et al.*, 1977) and rat liver membranes (Oldberg *et al.*, 1979) indicate that 4–5 heparan sulphate chains are bound to a protein core of molecular weight of 17,000–25,000 (Oldberg *et al.*, 1979).

It would appear that heparan sulphate and heparin proteoglycans show diverse characteristics when compared on a macromolecular level. The former has a relatively full complement of amino acids and may exist in lower molecular weight forms.

Various investigators have presented evidence suggesting that the bulk of intracellular heparan sulphate (molecular weight 6,000–12,000) is a degradation product of internalized heparan sulphate (Kraemer and Smith, 1974; Kresse *et al.*, 1975). While the mechanism of degradation is not clear, a number of enzymes have been demonstrated to have degrading activity towards heparan sulphate proteoglycan. These include proteolytic enzymes such as papain (Kraemer and Smith, 1974) (this activity may also be ascribed to the occurrence of random oxidation-reduction breaks due to the presence of cysteine-HCl, in an analogous manner to ascorbate action on heparin proteoglycans (Horner, 1971)) and pronase (Oldberg *et al.*, 1977; Johnston *et al.*, 1979). The existence of heparan sulphate depolymerizing endoglycosidases, to break down high molecular forms of heparan sulphate (in an analogous manner to the breakdown of macromolecular heparin to heparin chains, Section 4.3.1), have been isolated from rat liver lysosomal fractions (Höök *et al.*, 1975), platelets (Wasteson *et al.*, 1976), cultured human skin fibroblasts (both endoglucosaminidase and endohexuronidases were invoked, Klein *et al.*, 1976) and human placenta (Klein and von Figura, 1976). The fragments formed appear to be substrates for exoglycosidases and sulphatases.

4.3.3 Biosynthesis of protein core and carbohydrate linkage region

4.3.3.1 FORMATION OF PROTEIN CORE

It is probable that the first step associated with heparin or heparan sulphate multi-chain synthesis is the formation of the protein core. Its formation is poorly understood. Recent work by Robinson and Lindahl (1979) have shown that cycloheximide inhibits heparin proteoglycan synthesis (cf. inhibition also in chondroitin sulphate proteoglycan in cartilage (Telser *et al.*, 1965)). Furthermore, β-xyloside relieved this inhibition (in terms of heparin chain formation) but to a limited extent only, whereas the same compound relieves cycloheximide inhibition of chondroitin sulphate synthesis. The results for chondroitin sulphate biosynthesis have been interpreted in terms that protein synthesis necessarily precedes polysaccharide formation in the formation of chondroitin sulphate proteoglycan. β-xyloside relieves inhibition of polysaccharide synthesis by substituting for the serine–xylosyl residue of the protein core as an acceptor for chain elongation.

4.3.3.2 CHAIN INITIATION

It is assumed that the first enzyme of heparin chain formation is a xylosyltransferase, catalyzing the transfer of xylosyl groups from UDP–D-xylose to a core protein. Similar mechanisms have been invoked for chain initiation for other connective tissue glycosaminoglycans (for review see Rodén and Schwartz, 1975). An enzyme of this type has been discovered by Grebner *et al.* (1966ab) who showed that an extract of a mouse mastocytoma catalyzed the transfer of xylose to endogenous acceptors present in the crude tissue preparation. Information concerning this reaction with heparin formation is lacking. A recent study (Robinson and Lindahl, 1979) has shown that β-benzyl-D-xyloside is not as good an initiator for heparin chain biosynthesis as compared to chondroitin sulphate biosynthesis, which may suggest a different mechanism associated with its chain growth. The possibility that a major portion of the heparin chain, peripheral to the linkage region, is built on a lipid primer and therefore requiring the possible involvement of lipid intermediate in heparin biosynthesis cannot be discounted. After completion, such a chain could be transferred onto the poly(–ser–gly–) acceptor in which the serine residues are substituted with the linkage region. In view of the fact that other polysaccharides, including chondroitin sulphate, are also present in the mastocytoma. Rodén and Horowitz (1978) point to the possibility that one and the same xyloslytransferase may be responsible for initiation of all polysaccharides that are bound by a xylose–serine linkage.

4.3.3.3 COMPLETION OF THE POLYSACCHARIDE–PROTEIN LINKAGE REGION

The transfer of galactosyl and glucuronosyl residues to carbohydrate–serine compounds has been demonstrated. Each reaction represents the incorporation of a specific monosaccharide into the heparin–protein linkage region (Helting, 1971, 1972) by a specific glycosyltransferase.

The completed linkage structure –GlcUA–Gal–Gal–Xyl–Ser– is assembled by four different glycosyl-transferases, namely one xylosyl-transferase, two galactosyl-transferases and one glucuronosyl-transferase.

4.3.4 Chain–chain interaction

A number of polysaccharides that constitute the extracellular coats in bacteria or in extracellular matrices in algae and higher plants, are able to form gels by co-operative interaction between individual chains (see Rees, 1975 and references therein). The fact that chain–chain interactions may occur among connective tissue glycosaminoglycans in aqueous solution was first shown by Fransson in 1976. He found that polysaccharides (referred to as copolymeric galactosaminoglycans as in dermatan sulphate) that contained proportions of L-iduronic acid and D-glucuronic acid in approximately equal amounts have a marked tendency to aggregate. The interaction was demonstrated by gel chromatograhy and also by affinity chromatography using polysaccharide substituted gels. The chain–chain interactions occurred at physiological ionic strength and pH and did not require the presence of divalent ions. The interaction was dissociated with either 1 M urea, 0.5M guanidine HCl, 0.5M NaCl or 0.15M trichloroacetate, suggesting that both hydrogen bonding and hydrophobic interactions play a role in binding. Further, it was shown that the copolymeric glucosaminoglycans, heparin and heparan sulphate, together with chondroitin 4-sulphate were able to bind to dermatan sulphate-substituted agarose in the presence of 0.15M NaCl, whereas chondroitin 6-sulphate, hyaluronate and keratan sulphate did not. The chemical structure of the aggregating units in dermatan sulphate has been examined by subjecting the aggregating chains to various enzymic and chemical degradations prior to affinity chromatography on dermatan sulphate-substituted agarose (Fransson and Cöster, 1979). They proposed that the contact regions of aggregating dermatan sulphate chains contain 'mixed' sequences of the form [(glucuronic acid-N-acetylgalactosamine sulphate)$_m$ (-iduronic acid-N-acetylgalactosamine sulphate)$_n$]$_p$ where $p \geqslant 2.5$ and alternating sequences of $m=n=1$ and possibly mixed sequences where $m=2$.

In a recent report, Fransson (1977) has shown that heparan sulphate preparations with L-iduronic acid/D-glucuronic acid molar ratios between 0.4 and 0.7 and N-acetyl/N-sulphate molar ratios between 0.5 and 0.7 displayed chain–chain interaction as measured by gel chromatography.

These types of interactions that may potentially occur with heparin-like polysaccharides and with other glycosaminoglycans are of great importance, particularly with respect to chain organization in proteoglycan structures and intermolecular interactions in cellular and tissue regions.

REFERENCES

Anderson, P., Hoffman, P. and Meyer, K., *J. Biol. Chem.* **240**, 156 (1965).

Arbogast, B. S., Hopwood, J. J. and Dorfman, A., *Biochem. Biophys. Res. Commun.* **75**, 610 (1977).

Arnott, S. and Winter, W. T., *Fed. Proc.* **36**, 73 (1977).

Atkins, E. D. T. and Laurent, T. C., *Biochem. J.* **133**, 605 (1973).

Atkins, E. D. T. and Nieduszynski, I. A., in *Heparin: Structure, Function and Clinical Implications*, R. A. Bradshaw and S. Wessler (eds.), Plenum Press, New York, p19 (1975).

Atkins, E. D. T. and Nieduszynski, I. A., in *Heparin: Chemistry and Clinical Usage*, V. V. Kakkar and D. P. Thomas (eds.), Academic Press, London and New York, p. 21 (1976).

Atkins, E. D. T. and Nieduszynski, I. A., *Fed. Proc.* **36**, 78 (1977).

Atkins, E. D. T., Nieduszynski, I. A. and Horner, A. A., *Biochem. J.* **143**, 251 (1974).

Atkins, E. D. T. and Sheehan, J. K., *Nature New Biol.* **235**, 253 (1972).

Bansil, R., Yannas, I. V. and Stanley, H. E., *Biochim. Biophys. Acta* **541**, 535 (1978).

Barlow, G. H., Sanderson, N. D. and McNeill, P. D., *Arch. Biochem. Biophys.* **94**, 518 (1961).

Bettelheim, F. A., *Nature* **182**, 1301 (1958).

Bettelheim, F. A., *J. Phys. Chem.* **63**, 2069 (1959).

Bettelheim, F. A., in *Biological Polyelectrolytes*, A. Veis (ed.), Marcel Dekker, New York, p. 131 (1970).

Blout, E. R. and Stryer, L., *Proc. Natl. Acad. Sci. USA* **45**, 1591 (1959).

Branford-White, C. J., *Arch Oral. Biol.* **23**, 1141 (1978).

Branford-White, C. J. and Hardy, J. F., *FEBS Lett.* **76**, 229 (1977).

Braswell, E., *Biochim. Biophys. Acta* **158**, 103 (1968).

Burson, S. L., Fahrenbach, M. J., Frommhagen, L. H., Riccardi, B. A., Brown, R. A., Brockman, J. A., Lewry, H. V. and Stokstad, E. L. R., *J. Am. Chem. Soc.* **78**, 5874 (1956).

Cabassi, F., Casu, B. S. and Perlin, A. S., *Carbohydr. Res.* **63**, 1 (1978).

Casu, B., Torri, G. and Vercellotti, J. R., *Pharmacol. Res. Commun.* **11**, 297 (1979).

Chandrasekran, E. V., Spolter, L. and Marx, W., *Prep. Biochem.* **5**, 281 (1975).

Charles, A. F. and Scott, D. A., *Biochem. J.* **30**, 1927 (1936).

Chung, M. C. and Ellerton, N. F., *Biopolymers* **15**, 1409 (1976).

Elloway, H. F. and Atkins, E. D. T., *Biochem. J.* **161**, 495 (1977).

Fransson, L.-A., *Biochim. Biophys. Acta* **437**, 106 (1976).

Fransson, L.-Å., *Uppsala J. Med. Sci.* **82**, 130 (1977).

Fransson, L.-Å., *Carbohydr. Res.* **62**, 235 (1978).

Fransson, L.-Å. and Cöster, L., *Biochim. Biophys. Acta* **582**, 132 (1979).

Frannson, L.-Å., Huckerby, T. N. and Nieduszynski, I. A., *Biochem. J.* **175**, 299 (1978).

Gatti, G., Casu, B. and Perlin, A. S., *Biochem. Biophys. Res. Commun.* **85**, 14 (1978).

Grebner, E. E., Hall, C. W. and Neufeld, E. F., *Arch. Biochem. Biophys.* **116**, 391 (1966a).

Grebner, E. E., Hall, C. W. and Neufeld, E. F., *Biochem. Biophys. Res. Commun.* **22**, 672 (1966b).

Helting, T., *J. Biol. Chem.* **246**, 429 (1971).

Helting, T., *J. Biol. Chem.* **247**, 4325 (1972).

Hilborn, J. C. and Anastassiadis, P. A., *Anal. Biochem.* **39**, 88 (1971).

Hirano, S., *Life Sciences* **10**, 151 (1971).

Hirano, S., *Int. J. Biochem.* **3**, 677 (1972).

Höök, M., Pettersson, I. and Ögren, S., *Thromb. Res.* **10**, 857 (1977).

Höök, M., Wasteson, Å, and Oldberg, Å., *Biochem. Biophys. Res. Commun.* **67**, 1422 (1975).

Horner, A. A., *J. Biol. Chem.* **246**, 231 (1971).

Horner, A. A., in *Heparin:Chemistry and Clinical Usage*, V. V. Kakker and D. P. Thomas (eds.), Academic Press, London and New York, p. 34 (1976).

Horner, A. A., *Fed. Proc.* **36**, 35 (1977).

Jacobsson, I., Höök, M., Pettersson, I., Lindahl, U., Larm, O., Wirén, E. and von Figura, K., *Biochem. J.* **179**, 77 (1979).

Jamieson, A. M. and Maret, A. R., *Chem. Soc. Revs.* **2**, 325 (1973).

Jansson, L. and Lindahl, U., *Biochem. J.* **117**, 699 (1970).

Jansson, L., Ögren, S. and Lindahl, U., *Biochem. J.* **145**, 53 (1975).

Jaques, L. B. and McDuffie, N. M., *Semin, Thromb. Haemostas,* **4**, 277 (1978).

Johnson, E. A. and Mulloy, B., *Carbohydr. Res.* **51**, 119 (1976).

Johnston, L. S., Keller, K. L. and Keller, J. M., *Biochim. Biophys. Acta* **583**, 81 (1979).

Kavanagh, L. W. and Jaques, L. B., *Arzneim, Forsch.* **24**, 1942 (1974).

Klein, U., Kresse, H. and von Figura, K., *Biochem. Biophys. Res. Commun.* **69**, 158 (1976).

Klein, U. and von Figura, K., *Biochem. Biophys. Res. Commun.* **73**, 569 (1976).

Knecht, J., Cifonelli, J. A. and Dorfman, A., *J. Biol. Chem.* **242**, 6408 (1967).

Kosakai, M., Yamauchi, F. and Yosizawa, Z., *J. Biochem.* **83**, 1567 (1978).

Kraemer, P. M. and Smith, D. A., *Biochem. Biophys. Res. Commun.* **56**, 423 (1974).

Kresse, H., von Figura, K., Buddecke, E. and Fromme, H. G., *Hoppe-Seyler's Z. Physiol. Chem.* **356**, 929 (1975).

Lages, B. and Stivala, S. S., *Biopolymers* **12**, 127 (1973).

Lasker, S. E., *Fed. Proc.* **36**, 92 (1977).

Lasker, S. E. and Stivala, S. S., *Arch. Biochem. Biophys.* **115**, 360 (1966).

Laurent, T. C., *Arch. Biochem. Biophys.* **92**, 224 (1961).

Laurent, T. C., Tengblad, A., Thunberg, L., Höök, M. and Lindahl, U., *Biochem. J.* **175**, 691 (1978).

Liberti, P. A. and Stivala, S. S., *Arch. Biochem. Biophys.* **119**, 510 (1967).

Lindahl, U., *Biochim. Biophys. Acta* **130**, 368 (1966).

Lindahl, U., Cifonelli, J. A., Lindahl, B. and Rodén, L., *J. Biol. Chem.* **240**, 2817 (1965).

Lindahl, U. and Rodén, L., in *Glycoproteins*, A. Gottschalk (ed.), Elsevier, Amsterdam, p. 491 (1972).

Lloyd, A. G., Bloom, G. D. and Balazs, E. A., *Biochem. J.* **103**, 76P (1967).

LLoyd, A. G., Dodgson, K. S., Price, R. G. and Rose, F. A., *Biochim. Biophys. Acta* **46**, 108 (1961a).

Lloyd, A. G. and Dodgson, K. S., *Biochim. Biophys. Acta* **46**, 116 (1961b).

McDuffie, N. M., Dietrich, C. P. and Nader, H. B., *Biopolymers* **14**, 1473 (1975).

Mukherjee, D. C., Park, J. W. and Chakrabarti, B., *Arch. Biochem. Biophys.* **191**, 393 (1978).

Mutoh, S., Funakoshi, I. and Yamashina, I., *J. Biochem.* **84**, 483 (1978).

Nader, H. B., McDuffie, N. M. and Dietrich, C. P., *Biochem. Biophys. Res. Commun.* **57**, 488 (1974).

Nagarajan, M. and Rao, V. S. R., *Biopolymers* **18**, 1407 (1979).

Nieduszynski, I. A. and Atkins, E. D. T., *Biochem. J.* **135**, 729 (1973).

Nieduszynski, I. A. and Atkins, E. D. T., in *Structure of Fibrous Biopolymers* E. D. T. Atkins and A. Keller, (eds.), Colston Papers No. 26, Butterworths, London p. 323 (1975).

Nieduszynski, I. A., Gardner, K. H. and Atkins, E. D. T., in *Cellulose Chemistry and Technology*, Am. Chem. Symp. Series 48, p. 73 (1977).

Öbrink, B., Pertoft, H., Iverius, P.-H. and Laurent, T. C., *Connect. Tiss. Res.* **3**, 187 (1975).

Ögren, S. and Lindahl, U., *Biochem. J.* **125**, 1119 (1971).

Ögren, S. and Lindahl, U., *J. Biol. Chem.* **250**, 2690 (1975).

Oldberg, Å., Höök, M., Öbrink, B., Pertoft, H. and Rubin, K., *Biochem. J.* **164**, 75 (1977).

Oldberg, Å., Kjellén, L. and Höök, M., *J. Biol. Chem.* **254**, 8505 (1979).

Orr, S. F. D., *Biochim. Biophys. Acta* **14**, 173 (1954).

Park, J. W. and Chakrabarti, B., *Biochem. Biophys. Res. Commun.* **78**, 604 (1977).

Park, J. W. and Chakrabarti, B., *Biochim. Biophys. Acta* **544**, 667 (1978).

Patat, F. and Elias, H.-G., *Naturwissenschaften* **46**, 322 (1959).

Perlin, A. S., Casu, B., Sanderson, G. B. and Johnson, L. F., *Can. J. Chem.* **48**, 2260 (1970).

Perlin, A. S., *Fed. Proc.* **36**, 106 (1977).

Radhakrishnamurthy, B., Ruiz, H. A. Jr and Berenson, G. S., *J. Biol. Chem.* **252**, 4831 (1977).

Rees, D. A., in *MTP Intern. Rev. Sci. Biochemistry Ser. One, Biochemistry of Carbohydrates*, W. J. Whelan (ed.), University Park Press, Baltimore, Vol. 5, p. 1 (1975).

Righetti, P. G. and Gianazza, E., *Biochim. Biophys. Acta* **532**, 137 (1978).

Righetti, P. G., Brown, R. P. and Stone, A. L., *Biochim. Biophys. Acta* **542**, 232 (1978).

Robinson, H. C., Horner, A. A., Höök, M., Ogren, S. and Lindahl, U., *J. Biol. Chem.* **253**, 6687 (1978).

Robinson, H. C. and Lindahl, U. Personal communication.

Rodén, L. and Horowitz, M. I., in *Glycoconjugates, Mammalian Glycoproteins, Glycolipids and Proteoglycans*, M. I. Horowitz and W. Pigman (eds.), Academic Press, London and New York, Vol. 2, Section 1, p. 3 (1978).

Rodén, L. and Schwartz, N. B., in *MTP International Review of Science. Biochemistry of Carbohydrates*, W. J. Whelan (ed.) University Park Press, Baltimore, Vol. 5, p. 95 (1975).

Scott, J. E. and Tigwell, M. J., *Biochem. J.* **173**, 103 (1978).

Serafini-Fracassini, A., Branford-White, C. J. and Hunter, J. D., *FEBS Lett.* **32**, 116 (1973).

Spedding, H., *Adv. Carbohydr. Chem.* **19**, 23 (1964).

Stivala, S. S., *Fed. Proc.* **36**, 83 (1977).

Stivala, S. S. and Ehrlich, J., *Polymer* **15**, 197 (1974).

Stivala, S. S., Herbst, M., Kratky, O. and Pelz, I., *Arch. Biochem. Biophys.* **127**, 795 (1968).

Stivala, S. S. and Liberti, P., *Arch. Biochem. Biophys.* **122**, 40 (1967).

Stone, A. L., *Biopolymers* **3**, 617 (1965).

Stone, A. L., *Nature* **216**, 551 (1967).

Stone, A. L., in *Structure and Stability of Biological Macromolecules* (Biological Macromolecules Series, Vol. 2) G. Fasman and S. Timasheff (eds.), Marcel Dekker, New York, p. 353 (1969).

Stone, A. L., *Biopolymers* **10**, 739 (1971).

Stone, A. L., *Fed. Proc.* **36**, 101 (1977).

Stone, A. L. and Moss, H., *Biochim. Biophys. Acta* **136**, 56 (1967).

Sylvén, B. and Ambrose, E. J., *Biochim. Biophys. Acta* **18**, 587 (1955).

Telser, A., Robinson, H. C. and Dorfman, A., *Proc. Natl. Acad. Sci. USA* **54**, 912 (1965).

Wang, C. C. and Jaques, L. B., *Arzneim. Forsch.* **24**, 1945 (1974).

Wasteson, Å., Höök, M. and Westermark, B., *FEBS Lett.* **64**, 218 (1976).

Wolfrom, M. L., Weisblat, D. I., Karabinos, J. V., McNeely, W. H. and McLean, J. *J. Am. Chem. Soc.* **65**, 2077 (1943).

Wusteman, F. S., *Experientia* **28**, 887 (1972).

Yuan, L., *J. Polymer Sci. Symp.* **45**, 175 (1974).

Yuan, L., Dougherty, T. J. and Stivala, S. S., *J. Polym. Sci. A2*, **10**, 171 (1972).

Yuan, L. and Stivala, S. S., *Biopolymers* **11**, 2079 (1972).

Yurt, R. W., Leid, R. W., Jr, and Austen, K. F., *J. Biol. Chem.* **252**, 518 (1977).

Section C
Function

CHAPTER 5

Micro-ion and macro-ion interactions with heparin and related polysaccharides in model systems

5.1 MICRO-ION INTERACTIONS

The study of micro-ion interaction with glycosaminoglycans generally follows the interpretation of the polyelectrolyte properties of these polymers (for review see Comper and Laurent, 1978). The primary interaction is electrostatic in nature between the small charge of the mobile micro-ion and the electrostatic potential envelope of the polyelectrolyte.

This interaction forms the basis for a variety of properties exhibited by these polysaccharides, including ion distributions, osmotic pressure, ion transport, and electric potentials in polysaccharide compartments.

Other types of micro-ion interaction with charged polysaccharides have been distinguished; for example, specific stereochemical complexes between micro-ion and the hydroxyl groups of neutral sugars and sugar anions are known to occur (for reviews see Kretsinger and Nelson, 1976; Rees, 1975). This type of interaction is possible when a set of two or three hydroxyl groups is arranged stereochemically to fit into the co-ordination sphere of the cation with the displacement of water molecules from the hydration shell. The interaction, therefore, depends on the size and preferred co-ordination geometry of the cation. This type of binding is particularly prevalent with calcium, both in solid state and in solution but has not yet been shown to occur for the anionic glycosaminoglycans of vertebrate tissues.

In view of the highly anionic charge nature of the heparin-like polysaccharides, numerous studies have been performed on the interaction of small non-physiological cations with these polyions, which essentially reflect a charge interaction. In general, these interactions 1) dissociate at high net salt concentrations and with increasing concentration of a more highly charged binding competitor, and 2) are dependent on the nature of the interacting ion charges and concentration of interacting species. Interactions of this type include interactions of heparin-like polysaccharides with drugs and antibiotics (Blumenkrantz and Asboe-Hansen, 1974; Menozzi and Arcamone, 1978; for review Colburn, 1976) and multivalent cations, $Co(NH_3)_6^{+3}$ (Mathews, 1964) and Cu^{2+} (for review Stivala, 1977). Quaternary ammonium complexes and organic cation detergents have been utilized as effective precipitants for isolation and fractionation of glycosaminoglycans including heparin-like polysaccharides (for review Scott, 1973) and for fractionation by phase separation (Jennings and Hurst, 1974; Hurst *et al.*, 1978) in organic solvents. The induced metachromasia of dyes interacting with glycosaminoglycans has provided a large body of work in terms of glycosaminoglycan identification and semi-quantification in histological studies on tissues and in isolation and fractionation (Whiteman, 1973; for reviews Scott, 1973; Jaques, 1977). Interactions of glycosaminoglycans with metachromatic dyes has been reviewed by Phillips (1970). Interactions of glycosaminoglycans with fluorescent probes have involved studies with ethidium bromide (Chaudhuri *et al.*, 1975), berberine sulphate (Enerbäck, 1974) and acridine orange (for example Edwards *et al.*, 1978; Menter *et al.*, 1979).

5.1.1 Early studies

There is a paucity of information regarding the interaction of physiological micro-ions, such as Na^+, K^{2+} and Ca^{2+}, with the heparin-like polysaccharides. Most experiments have been performed *in vitro* under equilibrium conditions, and interpreted in a rather unsatisfactory, empirical fashion.

The studies by Salminen (1962) on the diffusion of K^+ and Na^+ ions from heparin solution into water resulted in an apparent increase in the Na^+/K^+ ratio in the diffusate as compared to systems without heparin, suggesting that heparin is binding K^+ in preference to Na^+. Similar conclusions were reached from equilibrium dialysis studies (Salminen and Luomanmäki, 1963). The expressions used in these studies for 'bound ions' were made in terms of an empirical concentration ratio-mass action law equation which has no physical basis and may be misleading. Indeed, Dorabialska and Plonka (1965), in their studies of sodium and caesium transfer rate coefficients in heparin, concluded that in high sodium concentration solutions heparin polymers have a relatively greater affinity for caesium ions, while in high caesium concentration solutions they have a relatively greater affinity for sodium ions. It follows that variation of the ion exchange equilibrium quotient is governed by its direct dependence on the concentration ratio of cations. Similar objections (see below) can also be assigned to the use of parameters describing the empirical 'equilibrium selectivity coefficients' in mixed micro-ion systems as used by Dunstone (1962), and others. Notwithstanding these considerations, the solution studies of Dunstone (1962) qualitatively reveal that bivalent cations, such as calcium and strontium, 'bind' strongly to the heparin-like polysaccharides as compared to mono-valent counterions.

5.1.2 Single counterion systems

The recent work of Manning (1974 and earlier references) seems to provide the most satisfactory theoretical framework to analyse the micro-ion interactions with polyions. Manning has developed expressions to describe the thermodynamic and transport properties of micro-ions in polyelectrolyte solutions directly in terms of the structure of the polyelectrolyte. The theory is developed for a state of infinite dilution in which the polyion molecules are infinitely far apart so they can be represented by infinitely long continuous charge. The relationships derived, therefore, are considered as 'limiting laws' although excellent agreement with theory has been found on studies with real solutions. The charged groups on the polyelectrolyte are seen as being screened by micro-ions from distinct groups on the same chain; thus, electrostatic contributions to the free energy are determined by chain segments that are short relative to the total chain length but may be long

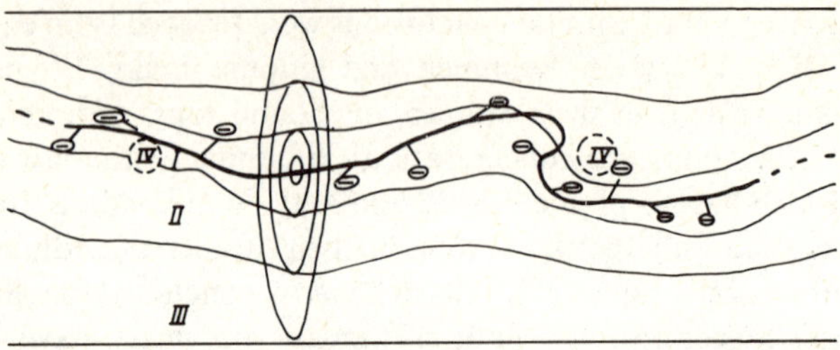

FIGURE 5.1 A simplified view of the various interactions envisaged to occur between micro-ions and polyion over a finite polyion length. I represents the phase of condensed counterions on the polyion; II represents the phase of uncondensed counter-ions interacting with the polyion-condensed micro-ion unit; III represents regions of micro-ion–macro-ion interactions; and IV represents sites of specific binding or 'site' binding. (From Comper and Laurent, 1978; reproduced with permission of the publishers.)

with respect to the dimensions of a small ion. This description is depicted in Figure 5.1, where a segment of the polyelectrolyte chain is viewed with its anionic charged groups distributed along it. There envisaged four types of interaction of the micro-ions in the vicinity of this polyelectrolyte segment. A brief description of these interactions follows. In assembling counterions onto the polyelectrolyte it can be demonstrated that a critical charge density ($\xi_{crit.}$) exists on the cylindrical polyion in infinitely dilute solution at which the counterions will associate onto the surface of the cylinder so that the net charge density (inclusive of the charge of the counterions) is maintained at a critical value. The charge-density parameter ξ, which forms a central part of Manning's theory, is defined by

$$\xi = e\beta/DkT \qquad 5.1$$

where β is the charge contained in a unit length of cylinder, e is the charge of an electron, D is the dielectric constant of the bulk solvent, k is the Boltzmann constant, and T is the temperature. Furthermore, the uniform charge density β is given by

$$\beta = Z_p e/b \qquad 5.2$$

where $b = L/P$ and L is the contour length of polyion that carries P charged groups of valence Z_p. Note that β and therefore ξ can be calculated from the known structure of the particular polyelectrolyte under study. The value of $\xi_{crit.}$ for monovalent charged groups ($Z_p = 1$), as in heparin-like polysaccharides, is

$$\xi_{crit.} = |Z_i|^{-1} \qquad 5.3$$

where Z_i is the valence of the counterion. For monovalent mobile ions ($Z_i = 1$) we have

$$\xi_{crit.} = 1$$

Thus for water as solvent at 37°C ($D = 74$), $\xi_{crit.} = 1$, corresponds to a charge spacing of $b = 0.724$ nm.

For the general case, if initially $\xi > \xi_{crit.}$, then the counterions will condense onto the polyelectrolyte chain until $\xi = \xi_{crit.}$; this defines interaction I in Figure 5.1. The condensed counterions have little or no translation mobility relative to the polyelectrolyte. Subsequent to the condensation of counterions onto the polyelectrolyte surface there is a high residual field due to the net charge of the polyelectrolyte and this affects the uncondensed or 'free' ions so that their behaviour is modified. For single counterion systems, the concentration of uncondensed counterions is $(\xi Z_i)^{-1} n_p$, where n_p is the concentration of polyion in equiv/L. The interaction of the uncondensed counterions with the polyelectrolyte (interaction II in Figure 5.1) is described by the Debye–Hückel mode of interaction in the limiting case. If $\xi < 1$, no counterion condensation occurs and all small ions interact with the fully dissociated polyion by Debye–Hückel forces. This type of interaction is known to contribute differently, in a quantitative manner, to various types of measurements (i.e. osmotic pressure, conductivity, diffusion) and has lead to a misinterpretation of ion 'binding' when its effects have not been considered (Manning, 1974). An empirical extension of the Manning treatment to finite salt concentrations has been carried out by Wells (1973) by taking into account the interaction between mobile ions that normally occur in the absence of the polyion (interaction III in Figure 5.1). The question of rates of exchange between condensed and uncondensed counterions is unresolved. It is important to note that there is no implication in this treatment that condensed counterions are in the condensed state for long times or at discrete sites or involved in specific binding. The differences of interaction of various micro-ions with a polyanion in Manning's theory are based solely on their valence. The situation may be complicated in certain systems by the occurrence of 'site' binding or specific binding which may not be based wholely on an electrostatic process (interaction IV in Figure 5.1).

Manning has developed expressions for activity coefficients, osmotic coefficients, self diffusion coefficients, and ion distribution coefficients for both single and mixed (i.e. mono- and di-valent counterions) systems all in terms of basic structural features of the polyion and the charge-density parameter ξ.

For heparin, the absolute line charge density will be approximately 3.5 charges per 0.86 nm or $b=0.245$ nm, which will correspond to a value of $\xi=2.96$ at 37°C. For monovalent counterion systems, the degree of condensation per mol of polyanion will be 0.66, which demonstrates that only one third of the heparin monovalent counterions are active and dissociated from the polyion under physiological conditions. For heparan sulphate, a charge density arbitrarily chosen as 2 charges per 0.9 nm (this will depend on the degree of sulphation) or $b=0.45$ nm with $\xi=1.61$ corresponds to a degree of condensation per mol of polyanion of 0.38. Values of intercharge distances for heparin-like polysaccharides calculated from various measurements, on the basis of the Manning theory, are given in Table 5.1. For self diffusion measurements there is very good agreement between calculated and experimental values of b for both heparin and heparan sulphate preparations. For reasons which are not clear, higher values of b are obtained by membrane electrode measurements.

It is worthwhile to note that differences in apparent 'binding' of monovalent ions to glycosaminoglycans with varying proportions of carboxyl and sulphate anionic groups have been observed. These differences have been ascribed to, in part, the intrinsic differences in the binding behaviour of the different anionic

TABLE 5.1
Interchange Distances of Heparin-like Polysaccharides as calculated from the Manning Theory.

Sample	Estimated total Charge Density (equiv/g) (see Footnote 5.1)	Calculated b (nm)	Method	Reference
Commercial heparin	N.I.	0.471[a]	Membrane electrode	Ascoli *et al.* (1961)
Commercial heparin	N.I.	0.262[b]	Self diffusion	Dorabialska and Plonka (1965)
Commercial hog mucosal heparin	5.44×10^{-3} [c]	0.238	Self diffusion	Ander *et al.* (1978)
Commercial hog mucosal heparin	4.48×10^{-3} [d]	0.430	Membrane electrode	Diakun *et al.* (1978)
	5.61×10^{-3}	0.291		
Commercial heparan sulphate	N.I.	0.558[b]	Self diffusion	Dorabialska and Plonka (1965)

N.I. = Not identified.
[a] Calculated from data given through equation $\gamma_{Na+} = \xi^{-1}(Z_i)^{-1}e^{-1/2}(\xi > 1)$ (Manning, 1969a) where γ_{Na+} = sodium ion activity coefficient.
[b] Calculated from data given through equation $D/D_0 = 0.87\ \xi^{-1}\ (\xi > 1)$ (Manning, 1969b) where D and D_0 are sodium ion diffusion coefficients in polyion solution and water respectively.
[c] Determined by titration.
[d] Determined by spectrofluorimetric titration with Acridine Orange (see also Footnote 5.1).

FOOTNOTE 5.1: A number of methods, apart from chemical analysis, have been published concerning the estimation of the line charge density of heparin-like polysaccharides and the ratio of total sulphate/carboxyl groups. Potentiometric procedures for total charge estimation have been reported by a conventional titration method (Kuettner and Lindenbaum, 1964) and by a conductimetric method (Casu and Gennaro, 1975). Estimates of the total sulphate/carboxyl ratio requiring sharp inflexion points of both charged groups have been obtained by titration using 70–80% vol. aq. dioxane as solvent (Kuettner and Lindenbaum, 1965) and by conductimetric method using addition of acid for the carboxyl inflexion point and addition of alkali for the sulphate groups (Casu and Gennaro, 1975). In the latter study, inflexion points for N-sulphate groups and O-sulphate groups were virtually identical. Furthermore, estimates of the carboxyl and sulphate groups were in good agreement with chemical analysis (with the exception of the carbazole–borate method for uronic acid carboxyl groups, see Section 2.2.2).

Titration methods, utilizing the spectral properties of metachromatic or fluorescent dyes, for the determination of polysaccharide line charge density have been employed. However, these measurements have not been confirmed by other methods (such as chemical analysis). It remains to be determined whether dye–polyanion interaction, for 1:1 stoichiometry between dye and polyanion charge, introduces varying fluorescent and metachromatic properties of the dye dependent on its type of interaction with the polyion as described in Figure 5.1 for micro-ion/polyanion interactions.

moieties of the charged polysaccharides, but as yet there is surprisingly little evidence to support this concept. A detailed comparative study of these 'potential' differences has yet to be performed and other important parameters such as anionic line charge density and charge distribution require consideration (see also Section 5.2.1). The Manning theory, in fact, does not distinguish between differences in the types of anionic charges of the polyion, but is based purely on valence and line charge density. Studies on chondroitin sulphate–keratan sulphate proteoglycans and other polyions indeed showed that their micro–ion interaction behaviour is not dependent on the carboxylate/sulphate ratio of the polyion (Comper and Preston, 1974, 1975). There is no support for the preferential binding of sodium to the carboxylate moieties of heparin as determined by the pH linewidth dependence of Na^{23}-nmr (Herwats *et al.*, 1977) unless the effect of line charge density with pH is also taken into account.

Differences associated with the binding of different monovalent ions to heparin-like polysaccharides and other polyanions have yet to be explained. However, recent studies by Joshi and Kwak (1978) on the interaction with alkali metal ions with dextran-sulphate suggests that the differences in the activity coefficients of the counterions at high concentrations of polyion can be adequately described by a model taking into account the differences in hydration of the counterion.

Although there have been some studies directed towards the binding of multivalent cations to heparin, including Ca^{2+}, there have been a number of criticisms levelled at the type of interpretation placed on these data. These interpretations of 'binding' have been based on empirically derived selectivity coefficients or distribution coefficients defined by the mass-action law but which have no physical basis. The errors in the interpretation of the distribution of the multivalent cations around a polyion lie, in part, in the neglect of the Debye–Hückel free energy of interaction of the polyion with the various counterions (Manning, 1974) and the estimation of the concentrations of the various species involved in the equilibrium interactions of the system (Marinsky, 1976). Interpretation of the Ca^{2+} binding to heparin as studied by self diffusion experiments (Ander *et al.*, 1978) were in accord with the

Manning theory, which suggests that although the binding of Ca^{2+} is high, it is not highly specific.

The relatively strong binding of divalent cations is equated with contraction of the molecule, presumably due to an increased binding of counterions with a concomitant decrease in effective charge density. This is apparent in the decrease in intrinsic viscosity of heparin (Chung and Ellerton, 1976), changes in the sedimentation and diffusion coefficients of heparan sulphate (Öbrink *et al.*, 1975) and X-ray diffraction (Section 4.1.1). Non-ideality of the heparin-like polysaccharides is reduced when Ca^{2+} is the counterion (Öbrink *et al.*, 1975).

5.1.3 Mixed counterion systems

Of great interest in relation to electrostatic micro–ion interaction to the heparin-like polysaccharides in the interaction for which two species of counterions of different valence are present, e.g. Ca^{2+} and Na^{+}. Analysis of the distribution of counterions surrounding a polyion in a mixed counterion system, particularly a system representing the physiological situation in which both Na^{+} and Ca^{2+} are present, has yet to be performed.

In view of the conformational changes that are known to occur with heparin on transformation of the counterion state from Na^{+} $\rightarrow Ca^{2+}$, a number of questions are immediately raised. What is the counterion state of heparin *in vivo*? Do changes in its counterion state and thus associated conformational changes amplify some other type of reaction in which heparin is involved? If there are fluctuations in counterion type, does this introduce additional fluctuations which again may be amplified in some other form of interaction? What is the counterion state of the heparin-like polysaccharides on the surface of cell plasma membranes and how will it affect the interaction properties of these polysaccharides with other macro-ions? The prediction of micro-ion and poly-ion organization *in vivo* is yet unattainable but such factors will be dependent on the polyelectrolyte concentration, the micro-ion concentration and their relative flux at the site of interaction.

Interest is also created in the mixed counterion system containing divalent cations by the work on an analogous system of the block type structures of alginates. Alginate is a binary heteropolymer containing 1,4 linked β-D-mannuronic acid and α-L-guluronic acid residues arranged along the chain in homopolymeric 'blocks' or of the two acid residues together with blocks of an alternating sequence. The similarity of this type of structural organization to that which may exist particularly in heparan sulphates has already been made. The micro–ion interaction of the various alginate structures shows marked differences dependent on the type of micro-ion and block structure. In particular, calcium can induce the alginate to gel. It is suggested that the homopolymeric blocks of L-guluronic acid are responsible for the formation of junctions, bridged by the Ca^{2+} ions in the gels, to give it rigidity; although reconstitution experiments with alginates and Ca^{2+} give gels with relatively poor rigidity. It seems other factors are also associated with gel induction and stability by Ca^{2+} *in vivo* (for review see Andresen *et al.*, 1977). Any evidence for such specific behaviour of micro-ion interaction with the possible block type structures in heparin and heparan sulphate has yet to be forthcoming.

Some attempts have been made to theoretically predict the counterion form of a polyion when in the presence of a Na^{+}–Ca^{2+} mixed ion system. Manning (1974) distinguishes a number of cases for the variation of the charge density parameter for salt added systems. For the case of $\xi > 1$, which holds for the heparin-like polysaccharides, the criterion for condensation of either Na^{+} or Ca^{2+} is satisfied. However, according to Manning (1974) the probability that Ca^{2+} will condense is infinitely greater than the corresponding probability for Na^{+}. Therefore, if there are sufficient Ca^{2+} counterions, they will condense preferentially onto the polyanion until the net value of ξ is lowered to 1/2 (see Eq. 5.3). In physiological terms, the degree of uptake of Ca^{2+} as condensed counterions, will depend on the ambient Ca^{2+} concentration and polyanion concentration. When ξ is lowered to 1/2, the excess Ca^{2+} as well as Na^{+} will remain in solution. If there are only as many Ca^{2+} ions as are required to lower ξ_{net} to a value between 1/2 and 1, then all the Ca^{2+} will condense whereas all the Na^{+} will remain in solution (since $\xi_{net} < 1$). If condensation of all the Ca^{2+}

results in a value of ξ_{net} which is greater than unity, then sufficiently many Na^+ counterions will condense in addition to all the Ca^{2+}, to lower the value of ξ_{net} to unity. Of particular physiological interest are the last two conditions where there is not enough Ca^{2+} to make ξ_{net} equal to 1/2. In these cases all the Ca^{2+} ions will be condensed and be inactive. Under such conditions the polyelectrolyte is behaving as a calcium 'sink' in spite of the presence of large excesses of sodium. However, recent experimental data by Kwak and co-workers (1976) on poly-styrenesulphonate do not bear these predictions out. Thus, although measurements of the activity coefficient for sodium are in excellent agreement with limiting law predictions, the calcium activity coefficient values show decreases with increasing weight fraction on Na^+, but the decrease is gradual and at no point do the activity coefficients become zero or even close to zero.

Although studies on heparin-like polysaccharides have yet to confirm this pattern of condensation for poly-styrenesulphonates there are two studies of immediate importance in relation to Ca^{2+} binding in the presence of Na^+. Öbrink *et al.* (1975) have studied the sedimentation and diffusion coefficients of heparan sulphate from human aorta in buffers of physiological ionic strength containing either NaCl or $CaCl_2$. The results indicate that the molecule contracts in the presence of calcium, presumably due to an increased binding of counterions with a concomitant decrease in 'active' charge density. When the polysaccharide was dissolved in a salt solution, physiological both in ionic strength and in sodium to calcium ratio (i.e. Na/Ca molar ratio 30–75), it sedimented as the sodium salt. Lages and Stivala (1973) have also indicated that calcium at low concentrations does not affect the intrinsic viscosity, sedimentation coefficients and partial specific volume of heparin in neutral buffer (0.1M Tris pH 7.5) and therefore concluded that these metal ion interactions caused no major change in the apparently random coil conformation of heparin.

5.1.4 H^+-ion interaction

The equilibrium which exists between hydrogen ions and the connective tissue polysaccharides has been studied essentially in terms of the binding of the carboxyl group only. The binding of H^+ to the sulphate moieties, being strong acids with pK_a values near 2, has not been published.

The titration of the carboxyl groups has been described empirically by the equation of Katchalsky and Spitnik (1957), where the apparent dissociation constant, K_a is described as a function of pH and degree of dissociation (α) by the following equation in which n is a constant

$$pH = pK_a + n \log \frac{\alpha}{1-\alpha}$$

Values of pK_a for the carboxyl groups of heparin given in Table 5.2 have been measured by potentiometric techniques and by variation of circular dichroic ellipticity at 210 nm and 230 nm with pH. The apparent pK_a values are sensitive to the ionic strength and generally decrease with increasing ionic strength. There are a number of features of these data worthy of comment. It appears that with O-sulphation of the iduronic acid residue, the pK_a of the carboxyl changes to a small extent with the iduronic acid in the $^1C_4$ chair form, whereas much larger changes are seen on transformation of the iduronate residue from the $^1C_4$ chair form to either the $^4C_1$ or $^1S_3$ form (Fransson *et al.*, 1978). The iduronic acid in the $^4C_1$ chair form has significantly higher pK_a than the glucuronic acid in the $^4C_1$ chair form, in all the glycosaminoglycans studied (Mathews, 1961; Fransson *et al.*, 1978).

5.2 MACRO-ION INTERACTIONS

5.2.1 General considerations

In view of the highly anionic charged nature of the heparin-like polysaccharides, their binding to positively charged macromolecules, mostly proteins, through electrostatic interaction has been documented widely over the last 40 years (for review see Jaques, 1967). This binding is characterized by its sensitivity to pH and ionic strength. The bonds are formed at low ionic strengths and

TABLE 5.2

Apparent pK_a values of the carboxyl groups of heparin and other polysaccharides

Sugar Residue	Method	Apparent pK_a	Ionic Strength	Reference
Heparin				
$IdUASO_3^-$	P	3.95	0.15	Fransson *et al.* (1978)
$IdUASO_3^-$	CD	5.1–5.3	0	Park and Chakrabarti (1977, 1978)
$IdUASO_3^-$	P	~4.8	?	Danishefsky (1975)
IdUA ($^1C_4$ form)	P	4.25–4.5	0.15	Fransson *et al.* (1978)
IdUA ($^4C_1$ and $^1S_3$ form)	P	3.25	0.15	Fransson *et al.* (1978)
GlcUA	P	2.85	0.15	Fransson *et al.* (1978)
Chondroitin Sulphate				
GlcUA	P	4.34	0	Mathews (1961)
GlcUA	P	3.38–3.35	0.1	Mathews (1961)
Dermatan sulphate				
IdUA	P	4.97	0	Mathews (1961)
IdUA	P	3.93	0.1	Mathews (1961)
Hyaluronate				
GlcUA	P	3.21	extrapolated to high ionic (∞) strength or 0-degree of dissociation	Laurent (1957)

Abbreviations:
P = potentiometric
CD = circular dichroism

dissolved by increasing simple electrolyte concentration. The binding properties of heparin-like polysaccharides have been studied in many diverse systems including plasma proteins, cell membranes, extracellular matrix proteins, intracellular proteins and enzyme systems. While some of these interactions may have biological functional significance (which has yet to be shown), it is likely that many of the diverse systems studied have merely incidental meaning. It is clear that only those interactions which are formed in solution at physiological pH and ionic composition or in other biological ionic–pH states are of physiological significance. (Without pondering on the many miscellaneous interactions that have been published, only those interactions which have created major interest and been studied in detail will be discussed in Chapters 6 and 7.) These interactions include those with cell membranes and subcellular structures, factors in lipoprotein metabolism, platelets, complement factors and coagulation factors.

In view of the ubiquitous nature of heparin–protein interactions, it is surprising that in many systems little is known of the physicochemical interactions that take place.

According to the mixing conditions, the polyanion–polycation reaction mixture is considered to exist in one of three states 1) two immiscible liquid phases (coacervation), 2) liquid–insoluble complex equilibrium state, 3) homogeneous solution with a soluble complex. In all cases, the nature of the electrostatic interaction may involve both nonspecific (being independent of the chemical sequence of the molecule so that any length of the molecule may react in the same way) and specific (dependent on certain specific sequences within a molecule) interactions. The complexes formed in each phase may be in the form of an electroneutral salt resulting from a stoichiometric reaction or of a complex with excess charge. A great deal of early work on various polyanion–polycation complexes has been performed by Bungenberg de Jong and collaborators (1949). More recently, Veis (1970) has analyzed these systems on a theoretical basis. The following factors are known to be of importance in determining the state of complexes formed; ionic strength, pH, charge density of the polyions, the concentration and mixing ratio of the polyions, the molecular weight distribution of the polyions, the molecular shape and flexibility of the interacting species. Although little work has been performed in the analysis of these parameters with heparin-like polysaccharide systems, it is clear that studies in this direction will give insights into the formation of supramacromolecular structures of the cell involving heparin-like polysaccharides such as storage and secretory granules in mast cells (Section 6.2.1) and other cells (Section 6.3).

A further consequence of heparin–polycation complexes is the effect of heparin on enzymes through the formation of such complexes. However, little is known of the physicochemical forces operating on enzyme activity. Factors such as local pH, electrostatic potential and solvation are likely to be affected in the micro-environment of the enzyme while in the presence of a polyanion. Effects on substrate activity are also likely to occur. Heparin, together with other polyanions, has been shown to inhibit many different enzymes (for reviews Bernfeld, 1966; Elbein, 1974), although the mechanisms of action are not well understood.

In binding studies of glycosaminoglycans with proteins it is generally found that the order of strength of binding is heparin > heparan sulphate > dermatan sulphate > chondroitin 4-sulphate > chondroitin 6-sulphate > hyaluronic acid. Obviously, a number of exceptions to this order occur in particular glycosaminoglycan–protein interacting systems. The order of binding in the region from dermatan sulphate to chondroitin 6-sulphate has not been clearly established. In any case, from this binding order and other data two common conclusions are drawn. Firstly, it has been proposed that sulphate groups interact more strongly with cationic materials than carboxyl groups. At this stage, there is no evidence to support a distinction between interacting properties of sulphate and carboxyl groups on a macromolecular level. In most systems studied, the individual parameters that may contribute to binding have not been varied independently. For example, the order of strength of binding commonly seen for the glycosaminoglycans given above follows the decreasing charge density (or increasing charge separation) of the polymer chain. The carboxyl content/disaccharide is the same for all these glycosaminoglycans, with the exception of keratan sulphate, so that variation in charge is due to the degree of sulphation. There is no evidence to suggest that the carboxyl groups may behave differently from sulphate groups in these systems, as carboxylate polymers of similar charge density and distribution have not been studied. Similarly, differences between polysaccharides with N-sulphate and O-sulphate moieties can hardly be invoked at the present time. The conclusions that could possibly be drawn are that the glycosaminoglycan electrostatic-interacting properties are dependent on charge density and charge distribution along the polysaccharide chain.

Secondly, another commonly asserted binding characteristic of glycosaminoglycans is that iduronate-containing polymers, namely heparin, heparan sulphate and dermatan sulphate, tend to bind more strongly than the other glycosaminoglycans. However, evidence that dermatan sulphate, which nominally has the same charge density as the chondroitin sulphates, binds more strongly that the latter is often conflicting. It is well established that both iduronate containing glucosaminoglycans and galactosaminoglycans are susceptible to a high degree of microheterogeneity in polymer residue sequence owing to the nature of the polymer modification reactions associated with the formation of the iduronate residue (Section 3.2). As a result, the heterogeneous nature of the charge distribution in these polymers is likely to account, in part, in amplification of binding to proteins.

5.2.2 Model systems

5.2.2.1 CONFORMATIONAL CHANGES

Gelman and Blackwell have published a series of papers in which they have followed the interaction between connective tissue polysaccharides and synthetic basic polypeptides by circular dichroism (Gelman and Blackwell, 1973; for review – Blackwell *et al.*, 1977). The polysaccharides induce a change in the conformation of the polypeptides from that of a charged coil to that of an α-helix. The different polysaccharides interacted with different strength in the order hyaluronate < chondroitin 4-sulphate < heparan sulphate < chondroitin 6-sulphate < keratan sulphate < dermatan sulphate < heparin. The polypeptides interacted with different strengths in the order polyornithine ($M = 105{,}000$) < polylysine ($M = 100{,}000$) < polyarginine ($M = 65{,}000$) i.e. the interaction increases with increasing length and pK_a of the polypeptide chains. The strength of the interaction could be measured by the different amount of α-helix that was formed and the melting temperature at which the α-helix disappears. An increase in the ionic strength or decrease in pH also disrupts the structure (Schodt *et al.*, 1976). A different order in the binding strength of the glycosaminoglycans was obtained on the basis of the ionic strength at which the polylysine–glycosaminoglycan complex is dissociated, i.e. chondroitin 6-sulphate < hyaluronate < chondroitin 4-sulphate < heparin < dermatan sulphate (Schodt *et al.*, 1976). In this case, it appears that the degree of conformational change and the strength of binding of the glycosaminoglycans to the synthetic polypeptides are to some degree independent. The induced conformational changes on synthetic peptides by glycosaminoglycans has been extended to other protein models in which the charge density of the protein has been reduced by inclusion of nonbasic amino acids (Stone and Epstein, 1977). Heparin has been shown to promote the formation of a *β*-structure for poly-(lysine:tyrosine 1:1) and an α-helix for poly-(lysine:phenylalanine 4:1). The studies have revealed little specificity for heparin as these conformational effects can be mimicked by other polyanions (Stone, 1977). The important conclusion from these experiments is that heparin-like polysaccharides may induce conformational changes in proteins due to a charge interaction. The nature of the conformational change will depend, in part, on the nature of the charge-density and -distribution in the protein.

5.2.2.2 HEPARIN-LINKED SEPHAROSE

A major advance in the study of heparin (and other connective tissue polysaccharides) interactions with macro-ions has been gained through the use of heparin coupled to CNBr-activated Sepharose (Iverius, 1971). The heparin, while being fixed to a solid support, may experience interactions with other solutes in solution. Macro-ions that form stable complexes with heparin-linked Sepharose at relatively low ionic strength may be dissociated at higher ionic strengths. This technique has then been employed in many studies on fractionation and purification of macro-ions. Another important feature of this system is its resemblance to a model for receptor linked heparin-like polysaccharides on the external surface of all plasma membranes (Section 6.4).

The physicochemical properties of heparin when linked to Sepharose as compared to its behaviour in solution are not clearly understood. Studies on the interaction of heparin-linked Sepharose with antithrombin III have been shown to be in good agreement with solution studies of this interaction (Björk *et al.*, 1979).

Recently, attempts have been made to study a number of variable parameters associated with this system as a function of its interaction properties (Bengtsson and Olivecrona, 1977). These studies have highlighted certain problems associated with quantifying interactions in the system, together with interpreting interactions on a comparative basis from results published in different laboratories. A series of gels with different degrees of substitution was used. For heparin–Sepharose gels defined as 1/*n*, then a 1/*1* gel corresponds to 2 mg of heparin added/ml of gel; for 1/*10* gels, 0.2 mg heparin was added/ml of gel, etc. In another series of gels that was studied, a 1/*1* heparin–Sepharose gel was mixed with varying volumes of unsubstituted Sepharose. These gels were designated 1 + *n*. Therefore, a 1 + *4* gel represents a mixture of one part of 1/*1* heparin–Sepharose with 4 parts of unsubstituted Sepharose. The amount of heparin substituted into the 1/*n* gels unfortunately was not determined, but is believed to be approximately 50% of the added heparin (Iverius, 1971). The ionic strength required for the dissociation of bound lipoprotein lipase from the series of these variously substituted heparin-Sepharose gels is shown in Figure 5.2 and 5.3. In studies of Bengtsson and Olivecrona (1977), it was clearly evident that the 1 + *n* gels and 1/*n* gels demonstrated different binding affinities. With the 1 + *n* gels, the differences in net concentration required for dissociation of bound enzyme were small (Figure 5.2), whereas with the 1/*n* gels these differences were larger (Figure 5.3). For the experiments shown in Figure 5.3 the NaCl concentrations required for dissociation of half the bound enzyme was about 0.9M for 1/1 heparin–Sepharose, 0.75M for a 1/*10* gel and only 0.55M for a 1/*1,000* heparin–Sepharose. However, for 1 + *n* gels, which repre-

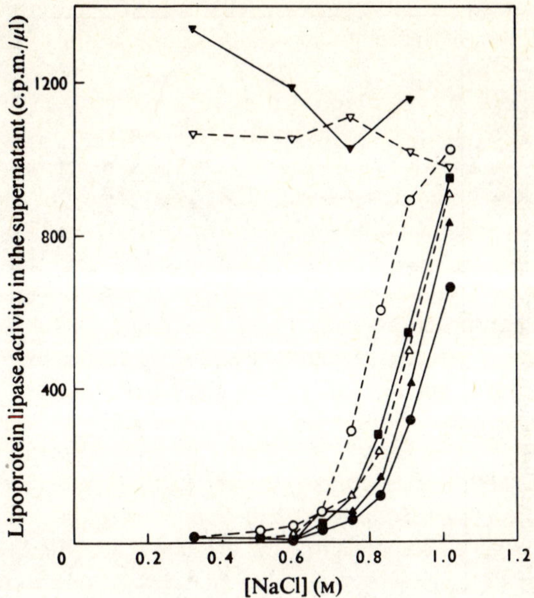

FIGURE 5.2 Displacement of lipoprotein lipase from $1+n$ heparin-Sepharoses by NaCl. Lipoprotein lipase (12 μg) was bound to the respective gels (0.1 ml) in a total volume of 1 ml. After 1 h the tubes were centrifuged at 2000 g for 2 min and samples of the supernatants were removed for assay of enzyme activity. Appropriate additions of 5MNaCl were then made, the contents of the tubes were mixed and incubated with gentle agitation for 20 min at 4°C. Then samples of the supernatants were taken after centrifugation, further additions of NaCl were made and the tubes incubated for a further 20 min and so forth. Thus all data for one heparin-Sepharose gel were obtained from the same tube. Altogether 170 μl of 5M-NaCl were added, causing a 15% dilution of the enzyme. Heparin-Sepharoses: ▲, 1 + *1*; △, 1 + *4*; ■, 1 + *9*; ●, 1/*1*; ○, 1/*10*; ▽, Ethanolamine-Sepharose; ▼, no gel added. (From Bengtsson and Olivecrona, 1977; reproduced with permission of the publishers.)

sents decreasing volumes of 1/1 gels, there was effectively no change in ionic strength required for dissociation. The $1/n$ gels showed decrease in ionic strength required for dissociation with decreasing degrees of heparin substitution in the gel phase. Whether this behaviour is peculiar to this interacting system is not known, but it does suggest some cooperative phenomenon associated with enzyme binding or a change in the properties of

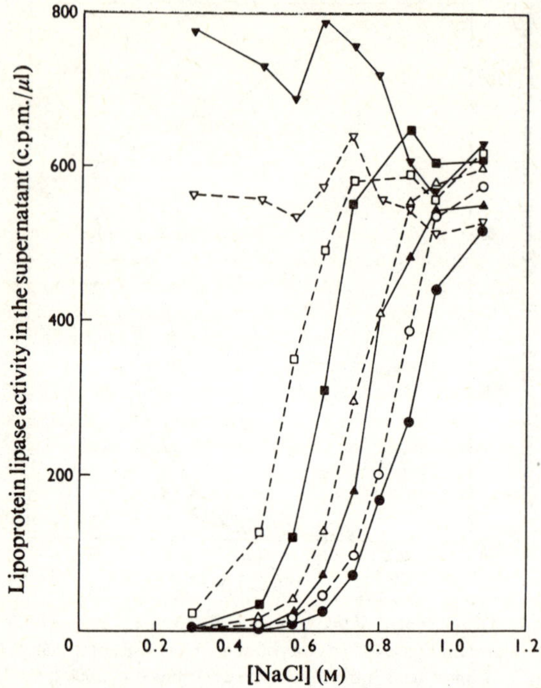

FIGURE 5.3 Displacement of lipoprotein lipase from l/n heparin-Sepharoses by NaCl. Lipoprotein lipase (7 μg) was bound to the respective gels (0.1 ml) in a total volume of 1 ml. The experiment was carried out as described for Figure 5.2. Altogether 200 μl of 5M-NaCl were added, causing a 20% dilution of the enzyme. Heparin-Sepharoses: ●, 1/*1*; ○, 1/*2*; ▲, 1/*5*; △, 1/*10*; ■, 1/*100*; □, 1/*1000*; ▽, Ethanol-amine-Sepharose; ▼, no gel added. (From Bengtsson and Olivecrona, 1977; reproduced with permission of the publishers.)

heparin on dilution in the gel phase. In any case, these studies do emphasize the fact that a comparison of the binding parameters of different macro-ions for heparin–Sepharose gels is probably only valid for gels with the same degrees of substitution. In view of the lack of detailed analysis of these gels used in published work a comparative analysis cannot be made, in most instances, at present.

REFERENCES

Ander, P., Gangi, G. and Kowblansky. A., *Macromolecules* **11**, 904 (1978).

Andresen, I. L., Skipnes, O., Smidsrød, O., Ostgaard, K. and Hemmer, P. C., in *Cellulose Chemistry and Technology*, Am. Chem. Soc. Symp. Series 48, p. 361 (1977).

Ascoli, F., Botré, C. and Liquori, A. M., *J. Phys. Chem.* **65**, 1991 (1961).

Bengtsson, G. and Olivecrona, T., *Biochem. J.* **167**, 109 (1977).

Bernfeld, P., in *The Amino Sugars*, E. A. Balazs and R. Jeanloz (eds.), Academic Press, London and New York, Vol. II, p. 213 (1966).

Björk, I., Danielsson, Å., Fish, W. W., Larsson, K., Lieden, K. and Nordenman, B., in *The Physiological Inhibitors of Coagulation and Fibrinolysis*, D. Collen, B. Wiman and M. Verstraete, (eds.), Elsevier/North Holland Biomedical Press, Amsterdam, p. 67 (1979).

Blackwell, J., Schodt, K. P. and Gelman, R. A., *Fed. Proc.* **36**, 98 (1977).

Blumenkrantz, N. and Asboe-Hansen, G., *Acta Pharmacol. Toxicol.* **34**, 27 (1974).

Bungenberg de Jong, H. G., in *Colloid Science*, H. R. Kruyt (ed.), Elsevier, Amsterdam, Vol. 2, p. 335 (1949).

Casu, B. and Gennaro, U., *Carbohydr. Res.* **39**, 168 (1975).

Chaudhuri, S., Phillips, G. O., Power, D. M. and Davies, J. V., *Int. J. Radiat. Biol.* **28**, 345 (1975).

Chung, M. C. M. and Ellerton, N. F., *Biopolymers* **15**, 1409 (1976).

Colburn, W. A., *Drug Metab. Rev.* **5**, 281 (1976).

Comper, W. D. and Laurent, T. C., *Physiol. Rev.* **58**, 255 (1978).

Comper, W. D. and Preston, B. N., *Biochem. J.* **143**, 1 (1974).

Comper, W. D. and Preston, B. N., *J. Coll. Int. Sci.* **53**, 391 (1975).

Danishefsky, I. in *Heparin: Structure, Function and Clinical Implications*, R. A. Bradshaw and S. Wessler (eds.), Plenum Press, New York (Adv. Exp. Med. Biol. Vol. 52) p. 105 (1975).

Diakun, G. P., Edwards, H. E., Wedlock, D. J., Allen, J. C. and Phillips, G. O., *Macromolecules* **11**, 1110 (1978).

Dorabialska, A. and Plonka, A., *Nukleonika* **10**, 331 (1965).

Dunstone, J. R., *Biochem. J.* **85**, 336 (1962).

Edwards, H. E., Diakun, G. P., Davies, J. V., Allen, J. C. and Phillips, G. O., *Biochem. Biophys. Res. Commun.* **85**, 1602 (1978).

Elbein, A. D., *Adv. Enzymol.* **40**, 29 (1974).

Enerbäck, L., *Histochem.* **42**, 301 (1974).

Fransson, L.-Å., Huckerby, T. N. and Nieduszynski, I. A., *Biochem. J.* **175**, 299 (1978).

Gelman, R. A. and Blackwell, J., *Arch. Biochem. Biophys.* **159**, 427 (1973).

Herwats, L., Laszlo, P. and Genard, P., *Nouv. J. Chim.* **1**, 173 (1977).

Hurst, R. E., Sheng, J. Y. P. and Ito, Y., *Anal. Biochem.* **85**, 230 (1978).

Iverius, P.-H., *Biochem. J.* **124**, 677 (1971).

Jaques, L. B., *Progr. Med. Chem.* **5**, 139 (1967).

Jaques, L. B., *Meth. Biochem. Anal.* **24**, 203 (1977).

Jennings, G. C. and Hurst, R. E., *Biochem. Biophys. Res. Commun.* **60**, 1209 (1974).

Joshi, Y. M. and Kwak, J. C. T., *Biophys. Chem.* **8**, 191 (1978).

Katchalsky, A. and Spitnik, P., *J. Polym. Sci.* **2**, 432 (1957).

Kretsinger, R. H. and Nelson, D. J., *Coord. Chem. Rev.* **18**, 29 (1976).

Kuettner, K. and Lindenbaum, A., *Science* **144**, 1228 (1964).

Kuettner, K. and Lindenbaum, A., *Biochim. Biophys. Acta* **101**, 223 (1965).

Kwak, J. C. T., Morrison, N. J., Spiro, E. J. and Iwasa, K. J., *Phys. Chem.* **80**, 2753 (1976).

Lages, B. and Stivala, S. S., *Biopolymers* **12**, 127 (1973).

Laurent, T. C., Thesis. Karolinska Institutet, Stockholm (1957).

Manning, G., *J. Chem. Phys.* **51**, 924 (1969a).

Manning, G., *J. Chem. Phys.* **51**, 934 (1969b).

Manning, G., in *Charged and Reactive Polymers. I. Polyelectrolytes*, E. Sélégny (ed.), D. Reidel Publishing Company, Dordrecht–Holland, p. 9 (1974).

Marinsky, J. A., *Coord. Chem. Rev.* **19**, 125 (1976).

Mathews, M. B., *Biochim. Biophys. Acta* **48**, 402 (1961).

Mathews, M. B., *Arch. Biochem. Biophys.* **104**, 394 (1964).

Menozzi, M. and Arcamone, F., *Biochem. Biophys. Res. Commun.* **80**, 313 (1978).

Menter, J. M., Hurst, R. E., Nakamura, N. and West, S. S., *Biopolymers* **18**, 493 (1979).

Öbrink, B., Pertoft, H., Iverius, P.-H. and Laurent, T. C., *Connect. Tiss. Res.* **3**, 187 (1975).

Park, J. W. and Chakrabarti, B., *Biochem. Biophys. Res. Commun.* **78**, 604 (1977).

Park, J. W. and Chakrabarti, B., *Biochim. Biophys. Acta* **544**, 667 (1978).

Phillips, G. O., in *Chemistry and Molecular Biology of the Intercellular Matrix*, E. A. Balazs (ed.), Academic Press, London, and New York, Vol. 2, p. 1033 (1970).

Rees, D. A., in *MTP Intern. Rev. Sci. Biochemistry Ser. One, Biochemistry of Carbohydrates*, W. J. Whelan (ed.), University Park Press, Baltimore, Vol. 5, p. 1 (1975).

Salminen, S., *Nature* **195**, 908 (1962).

Salminen, S. and Luomanmäki, K., *Biochim. Biophys. Acta* **69**, 533 (1963).

Schodt, K. P., Gelman, R. A. and Blackwell, J., *Biopolymers* **15**, 1965 (1976).

Scott, J. E., *Biochem. Soc. Trans.* **1**, 787 (1973).

Stivala, S. S., *Fed. Proc.* **36**, 83 (1977).

Stone, A. L., *Fed. Proc.* **36**, 101 (1977).

Stone, A. L. and Epstein, P., *Biochim. Biophys. Acta* **497**, 298 (1977).

Veis, A., in *Biological Polyelectrolytes*, A. Veis (ed.), Marcel Dekker, New York, p. 211 (1970).

Wells, J. D., *Biopolymers* **12**, 223 (1973).

Whiteman, P., *Biochem. J.* **131**, 343 (1973).

CHAPTER 6

Tissue distribution and cellular localization

6.1 TISSUE DISTRIBUTION

6.1.1 Vertebrates

Heparin is widely distributed in mammalian connective tissues and appears to be exclusively located, intracellularly, in the mast cells of these tissues. No evidence has been published to negate the possibility that this polysaccharide may naturally occur in other types of cell or in the extracellular matrix. The yields of heparin from its major sources are summarized in Table 6.1. Commercial sources of heparin have generally been confined to beef lung and hog intestinal mucosa.

Toledo and Dietrich (1977) have performed a comparative study of the tissue distribution of all the sulphated glycosaminoglycans, with the exception of keratan sulphate which was not analyzed, in five mammalian species viz., guinea-pig, rabbit, dog, hog and human. Identification of the various glycosaminoglycans was based on a combination of electrophoretic techniques, analysis of disaccharide products resulting from the action of specific degrading enzymes and anticoagulant activity. Determination of the glycosaminoglycan was primarily based on a toluidine blue dye-binding technique for glycosaminoglycans fractionated by gel electrophoresis (see Footnote 6.1). They established that heparin

FOOTNOTE 6.1

It is clear that a great deal of caution should attend studies which utilize metachromatic dyes for identifying and estimating glycosaminoglycans quantitatively both in solution, gels and dried films for histological studies. Several approaches have been used. These basically involve 1) the use of a particular dye to measure the absolute charge density of the polyanion by spectrophotometric titration and thereby obtain a measure of its concentration, 2) use of a calibrated distribution in tissues in mammalian species is species specific and that large variations are observed on a comparative basis. The heparin content of dog lung was relatively high, although their values for various tissues, particularly bovine lung, appear lower in comparison to yield values obtained by Linker and Hovingh (1973) (Table 6.1). Heparin was also identified in substantial amounts (i.e. $>30\ \mu g$ heparin/g dry tissue) in dog liver, guinea pig

scale relating metachromatic behaviour to polyanion concentration (normally obtained in aqueous solution) and 3) the use of the critical electrolyte principle of Scott (1973 and earlier references) which relates particular simple electrolyte concentrations required to dissociate metachromatic dyes from their substrate. The nature of this interaction appears to be dependent on both the type and density of charge on the polyanion (Scott, 1973).

The problems associated with these various techniques have been studied particularly with the use of Alcian blue, Safronin O and Toluidine blue O (Goldstein and Horobin, 1974ab; Tas, 1975, 1977), in the study of histological sections of various tissues, including mast cells, and in model connective tissue systems (i.e. polyacrylamide films containing entrapped glycosaminoglycans). A possible artifact that may be introduced by using a particular dye is its susceptibility to aggregate in aqueous solution and that this degree of aggregation is affected by temperature, salt concentration and the presence of organic solvents. The aggregation of dyes has been associated with the time dependence and irreversibility of staining various tissue sections (Stone, 1969; Goldstein and Horobin, 1974ab; Tas, 1975, 1977). Further problems arise in terms of gaining information on the substrate being studied. Most often, there is hardly any information concerning the nature of the chemical heterogeneity or molecular polydispersity of polyanions present and which may significantly alter the metachromatic behaviour of the dye. The same argument applies to metachromatic identification and electrophoretic behaviour of heparin-like polysaccharides in gels (Horner, 1975). Factors such as molecular weight of substrate (Scott and Willet, 1966; Morgan *et al.*, 1972), the concentration of substrate and the method of fixing (Tas, 1975) may all affect the metachromatic behaviour of the dye.

Conclusions concerning the state of the polyanionic charge in the tissue must be viewed in terms of the possibility of masking of charged groups by material with which the dye cannot displace, and that charges available for dye uptake can only be viewed in terms of competition of the dye with other counterions in the particular area. It is also clear that quantitative analysis of heparin in tissue samples by colourimetric assays is not particularly accurate (Section 2.2.2). For example, in studies on the binding of Acridine orange to heparin in mast cells, Lagunoff (1974) has used carbazole assay to determine the extent of binding per mol of disaccharide. This procedure may overestimate the uronic acid quantity (Table 2.1) and therefore underestimate the amount of Acridine orange bound so that quantitative conclusions concerning the state of heparin charge in mast cell granules is in doubt.

TABLE 6.1
Main Sources of Heparin-Like Polysaccharides

Source	Heparan Sulphate	Heparin	Reference
	(μg glycosaminoglycan/g dry de-fatted tissue)		
Bovine lung	534	410	Linker and Hovingh (1973)
Bovine lung	515	45	Toledo and Dietrich (1977)
Canine lung	525	346	Toledo and Dietrich (1977)
Human kidney	88	<1.0	Toledo and Dietrich (1977)
Canine liver	201	41	Toledo and Dietrich (1977)
Human skin	39	39	Toledo and Dietrich (1977)
Human umbilical cord	700	0	Cifonelli and King (1970)
Human aorta	≃4000	N.I.	Stevens *et al.* (1976)
Human aorta	1409	?	Toledo and Dietrich (1977)
Porcine ileum	353	112	Toledo and Dietrich (1977)
	(μg Uronic Acid/g dry de-fatted tissue)		
Adult feline brain	46–106	N.I.	Young and Custod (1972)
Non-human primate aortas	≃1500	N.I.	Radhakrishnamurthy *et al.* (1978b)
Bovine aorta	378	50	Radhakrishnamurthy *et al.* (1978a)
Canine stomach mucosa	N.I.	600	Kupchella and Steggerda (1972)
	(μg Hexosamine/g dry de-fatted tissue)		
Human coronary artery	410	N.I.	Tammi *et al.* (1978)

N.I. = Not identified.
Note: For yields of heparin-like polysaccharides/wet weight of tissue, see Brimacombe and Weber (1964).

and hog ileum and human skin (Toledo and Dietrich, 1977). The presence of heparin in human aorta was questionable considering the relatively low amounts of this polysaccharide (Toledo and Dietrich, 1977). Human sources of heparin in sufficient quantities for analysis have been confined to skin (Toledo and Dietrich, 1977) and lung (Horner, 1977). Horner (1977) has found heparin in the lung, liver and skin of Cynomolgus monkeys.

Another common source of heparin used for investigation is the lung and intestine of the finback whale (*Balaenoptera physalus*); this heparin exhibits high anticoagulant activity and is often designated ω-heparin (Table 1.1) (Yosizawa, 1964). No recovery yields for this material have been reported.

In contrast to the apparent spasmodic distribution of heparin in tissues of various animals, Toledo and Dietrich (1977) found that the distribution of other sulphated glycosaminoglycans, including heparan sulphate, exhibit the same types and proportions of glycosaminoglycans in the same tissue from different adult mammals. The absolute quantities of the glycosaminoglycans reported may, however, vary as evidenced with rat tissues (Dietrich *et al.*, 1976). Heparan sulphate is particularly widespread, and its yields from major sources are listed in Table 6.1. Heparan sulphate content is relatively high in the kidney (see also Murata, 1975), liver, ileum and lung; being in the range of 50–80% of the total sulphated glycosaminoglycan content of these tissues. The ubiquitous nature of heparan sulphate localization probably reflects its property as a cell surface membrane component (Section 6.4). Other sources of heparan sulphate (not listed in Table 6.1) include rabbit uterus (Endo and Yosizawa, 1976; Takata and Terayama, 1977), anuran skins (Hata and Nagai, 1973), bovine liver capsule (Serafini-Fracassini *et al.*, 1973), cornea (Meier and Hay, 1974; Hart, 1976), brain and its subcellular fractions (for review Margolis and Margolis, 1977), guinea pig renal medulla (Lis and Monis, 1978), rat salivary gland, larynx and uterus (Dietrich *et al.*, 1977), dentine (Branford-White, 1978b) and glomerular basement membrane (Kanwar and Farquhar, 1979).

Topographical distributions of heparin-like polysaccharides have been studied in beef lung (Seethanathan *et al.*, 1975; Wusteman, 1972), kidney (Murata, 1975; Lis and Monis, 1978), anuran skins (Hata and Nagai, 1973) and brain (for review, see Margolis and Margolis, 1977; Branford-White, 1978a). The results have shown that the heparin-like glycosaminoglycan distribution is topographically specific, although no conclusions can yet be drawn for the reasons for this apparent specificity. In general, the results reported for heparin-like polysaccharide in tissues have primarily focused at establishing whether these polysaccharides are present and their relative amounts as compared to other glycosaminoglycans. Detailed characterization and extensive fractionation of these materials await further investigation.

Numerous studies have shown a correlation of the content of heparin-like polysaccharides relative to other glycosaminoglycans in pathological states and aging of tissues. These considerations have been related particularly to heparan sulphate content and distribution; heparin has been studied only to a limited degree in this context. A review of changes associated with heparan sulphate in aging and pathological tissue states is beyond the scope of this work and the reader is referred to other works on this important subject (Neufeld *et al.*, 1975; Dorfman and Matalon, 1976; McKusick *et al.*, 1978).

6.1.2 Invertebrates

Clams have yielded heparin-like material with high blood anticoagulant activity. Thomas (1951, 1954) obtained such a substance from the common surf clam, *Spisula solidissima.* Anticoagulant substances from various species of clams (Burson *et al.*, 1956) have been reported to have molecular weights, optical rotations and anticoagulant activities which are higher than those obtained from mammalian heparins (see Table 2.1, Table 4.5). The term 'mactin' was given to these substances (Frommhagen *et al.*, 1953) to distinguish them from mammalian heparins. Frommhagen *et al.* (1953) extracted mactin A and B from two species of clams, *Mactra spisula* and *Artica islandica. In vitro* anticoagulant activities of the purified neutral sodium salts were 130–150 U.S.P. units/mg for mactin A and 150–180 U.S.P. units/mg for mactin B as compared with 120–145 U.S.P. units/mg for the heparin neutral sodium salt. Cifonelli and Mathews (1972) showed, on the basis of analysis of hexosamine-containing products formed after reaction with nit-

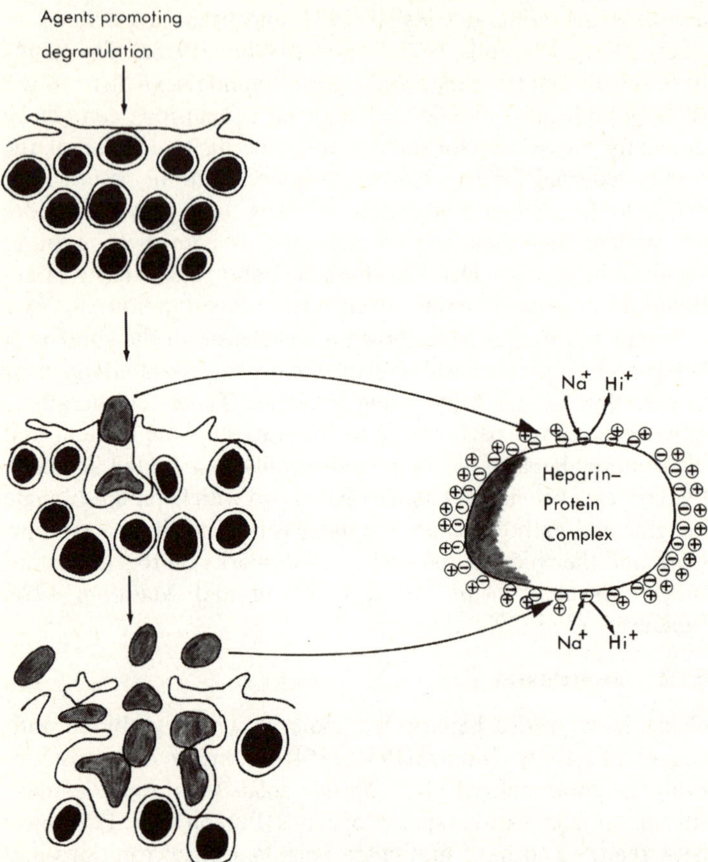

FIGURE 6.1 Schematic drawing to show sequential exocytosis of mast cell granules. For explanation see text. (Adapted from Uvnäs, 1977; reproduced with permission of the publishers.)

rous acid, that the mactins have molecular structures similar to mammalian heparins; although in contrast to the latter, the mactins appear to have larger proportions of glucuronic than of iduronic acid.

Rahemtulla and Løvtrup (1975) have isolated heparin from *Anodanta* (Bivalvia) with a molar ratio of N-sulphate:glucosamine of 1:2. They were unable to detect heparin-like material in other invertebrates studied, which included species belonging to the phyla *Annelida*, *Platyhelminthes* and *Aschelminthes*, although other sulphated glycosaminoglycans were present in varying proportions.

Cássaro and Dietrich (1977) have studied the sulphated glycosaminoglycan composition of 22 species of invertebrates belonging to the phyla *Arthropoda*, *Mollusca*, *Annelida*, *Tunicata*, *Echinodermata*, *Coelenterata* and *Porifera*. The sulphated glycosaminoglycans were identified and analysed quantitatively by a combination of agarose gel electrophoresis, enzymatic degradation with specific enzymes and dye-binding techniques. It was shown that all the species studied contain variable amounts of one or more types of sulphated glycosaminoglycans; most of which are similar to the ones found in vertebrates. Heparin sulphate appears to be present in relatively substantial amounts (30–90% of the total sulphated glycosaminoglycan) in phyla *Arthropoda*. All species of *Mollusca* studied (from subclasses *Filobranchia*, *Lumellibranchia* and *Coleoidea*) contain heparin-like polysaccharides to varying extents. It is noteworthy that the single sulphated glycosaminoglycan present in the two species studied from the subclass, *Lumellibranchia*, had anticoagulant activity of 150 IU/mg and is similar to mactins isolated from other species of this class (*vide supra*). The large amount of this material present in *Lumellibranchia* (i.e. 800–1,000 μg/g dry tissue) is about one order of magnitude higher than beef lung (the usual source of commercial heparin). Heparin-like polysaccharides were found in small amounts from one species of the phyla *Annelida* and two species of the phyla *Coelenterata*. Species from phyla *Tunicata*, *Echinodermata* and *Porifera* had no identifiable heparin-like polysaccharides. In contrast, Kinoshita (1971) has identified a heparin-like polysaccharide associated with the early development of the sea urchins *Clypeaster japonicus*, *Hemicentrotus pulcherrinius* and *Pseudocentrotus depressus* (phylum *Echinodermata*).

Except for hyaluronate, no other glycosaminoglycans have been reported in bacteria. Cássaro and Dietrich (1977) were unable to detect sulphated glycosaminoglycans in *Flavobacterium heparinum* and *Staphylococeus aureus*, in fungi *Blastocladiella emersonii* and *Pythium sp.* and in the protozoans *Trypanosoma cruzi*,

Acantamoeba sp. and *Tetrahymena pyriformis.* They suggest that the emergence of sulphated glycosaminoglycans corresponds to the emergence of tissue-organized life forms and that the sulphated glycosaminoglycans might be involved in the process of cell differentiation conferring to the cells some of their particular characteristics such as adhesiveness, recognition and possibly their morphology.

6.2 MAST CELL

Heparin is synthesized and then stored in basophilic granules of mast cells. Thus far, it is the only established source of endogenous heparin in vertebrates. The mast cells were first described by Ehrlich (1879) (cited in Wilhelm *et al.*, 1978) and are characterized as having large numbers of basophilic granules giving a metachromatic colour with dyes, which is due to the presence of sulphated glycosaminoglycans of which heparin is one. The accumulation of these granular cells in chronically inflamed tissues in passive venous congestion of the lung and in neoplastic tissue prompted Ehrlich to suggest that the mast cell arose from connective tissue cells that had been 'masted' (i.e. gorged with food so resulting in their high content of granules, hence the name Mastzella, mast cells).

As with heparin distribution in mammalian tissues (Section 6.1), mast cell content and localization vary greatly for various animals. Jaques (1975) has noted the particularly large quantities of mast cells associated with dog liver, and the subcutaneous tissue of the rat, although he claims they are not found in rabbit tissues. The significance of these peculiar distributions is not known. The distribution of mast cells in various tissues has been reviewed by Kiernan (1976) and Wilhelm *et al.* (1978). The mast cells are rarely found in blood. The actual origin of mast cells and their legitimacy as a uniform, distinct cell line also remain in doubt.

It is surprising that there is a paucity of carefully evaluated information on the identification of heparin in mast cells. Initially, the distribution of heparin in tissues was reported to correspond to the distribution of mast cells within the same tissues (Wilander,

1938; Jorpes, 1946). A large body of evidence for the purported presence of heparin in mast cells has come primarily from spectrochemical techniques. These include 1) the binding-induced metachromasia of Toluidine blue O and various other aniline dyes to the basophilic substrates of mast cell granules (for review see Jaques, 1975; Jaques *et al.*, 1977) and 2) the binding-induced fluorescence emission changes of berberine sulphate to granule polyanions (Enerbäck, 1974) as measured by cytofluorometric techniques (Berlin and Enerbäck, 1978 and references therein). Apart from the familiar but valid objections raised by using these histochemical techniques (see also Footnote 6.1), these techniques lack substantial proof in terms of glycosaminoglycan identification in these cells unless used in combination with other techniques.

Early attempts at chemical identification of heparin in normal mast cells were made by Schiller and Dorfman (1959) in using paper chromatography, anticoagulant activity, and hexosamine and uronic acid analysis. Lagunoff *et al.* (1964) extracted 90% of heparin from mast cells with 2M KCNS; the heparin analysis was based on N-sulphate hexosamine residues. More recently, Yurt *et al.* (1977a) have been able to produce [^{35}S]-heparin proteoglycan *in vitro* by incubation of highly purified rat peritoneal mast cells with [^{35}S]-sulphate and *in vivo* by injection of [^{35}S]-sulphate into rats. The heparin was identified by metachromasia, uronic acid, anticoagulant activity and antithrombin activity. Whether heparin occurs in other cells is not known. Heparin biosynthesis has been studied intensively in mouse mastocytoma (Section 3.2) although knowledge of its synthesis and turnover in normal mast cells is lacking. Exogenous or administered heparin is not taken up by the mast cells.

Mast cell numbers, as well as heparin content, are found to be strongly related to body weight and age of rats studied (Enerbäck and Mellblom, 1978). The heparin content, as measured by cytofluorimetric techniques, in the mast cell populations appeared to be either approximately normally distributed or slightly positively skewed. Mast cells vary in size and shape according to tissue and species of test animal (Wilhelm *et al.*, 1978). Large variations in both 5-hydroxy tryptamine and heparin content occur within granule populations of individual cells (Gustafsson and Enerbäck, 1978). There is as yet no information regarding the possibility of different types of heparin associated with mast cell heterogeneity and granule population heterogeneity. Mast cell function is, as yet, not clearly understood and may vary according to the tissue and its age.

There are a number of other cells, which like mast cells, show granule metachromasia when treated with certain aniline dyes but contain sulphated glycosaminoglycans other than heparin. Murata *et al.* (1974) have studied the glycosaminoglycans of bovine leucocytes, as identified by ion exchange columns, enzyme degradation and chromatographic techniques and demonstrated the presence of chondroitin sulphate, hyaluronate and heparan sulphate in the proportions of 75–80%, 10–15% and 10% respectively. Analysis of glycosaminoglycans (based on specific enzyme degradation) isolated from [^{35}S]-sulphate prelabelled purified granules of guinea pig basophilic leucocytes demonstrated that of the glycosaminoglycans synthesized, chondroitin sulphate and dermatan sulphate constituted the major portion together with small amounts of heparan sulphate (Orenstein *et al.*, 1978). Chondroitin sulphate has been identified in rabbit basophilic leucocytes (Sue and Jaques, 1974) and chicken mast cells (Sue and Jaques, 1976) together with the purported lack of heparin in these cells, as determined by an electrophoretic densitometry technique. Histochemical investigations of mucosal mast cells of rat small intestine failed to identify heparin (Tas and Bernsden, 1977). Although the major glycosaminoglycan of a murine mastocytoma is heparin, other sulphated glycosaminoglycans have been identified in smaller quantities (Chandrasekaran *et al.*, 1975; Hurst *et al.*, 1978).

Mast cells are numerous in the tissues of the rat, whereas basophils are reported to be infrequent and probably absent in the blood of that species. The converse is true for the rabbit, in which mast cells are infrequent in the mesentery and omentum of adults, whereas basophils quite often account for some 30% of circulating leucocytes. Both mast cell and basophil occur in the guinea pig and man (Jacques *et al.*, 1977).

6.2.1 Mast cell granules

Normal mast cell granules contain, in addition to heparin and possibly other glycosaminoglycans, biogenic amines and protein as described in Table 6.2. The histamine concentration in the granule is particularly high (0.3–1.2 M) and is thought to be bound to both granule protein and heparin by electrostatic interaction (see below). In rodent mast cells, 5-hydroxytryptamine (serotonin) also appears to be present. Yurt and Austen (1977) have shown recently that the mast cell chymase ($M = 25{,}000$ as determined by gel chromatography) (see also Everitt and Neurath, 1979) accounts for the major portion of the protein content in the granule. Another enzyme, β-glucosaminidase, also appears to be present (Lagunoff *et al.*, 1964) together with two less prominent basic proteins (Bergqvist *et al.*, 1971). The content of multi-valent metal cations within the granules has not received much attention although a value of 4 nequiv Zn/10^6 cells has been reported by Uvnäs *et al.* (1975) its presence is probably not important in terms of quantitative granule interactions, particularly histamine binding.

The heparin content of mast cells represents about 2% wt/cell volume or 30% of the total organic material of the granules (Lagunoff *et al.*, 1964). In view of a lack of detailed chemical analysis of heparin in mast cells, it is not possible to give an accurate estimate of its fixed negative charge concentration in granules. However, assumption of an equivalent weight of 5 mequiv/g for heparin will give a charge concentration within the granule in the range of 0.6–1.7 equiv/L (Table 6.2). The heparin in the granule has been shown to exist in its macromolecular form at least in rat peritoneal mast cells (Yurt *et al.*, 1977a), so in calculation of its molar granule concentration a molecular weight of 7.5×10^5 has been used (Table 6.2).

Little is known of the nature of the complex between heparin, protein and biogenic amines in the granule other than it exists as an insoluble ionic complex which can be solubilized at high salt concentrations. Indeed, the significance of such high concentrations of materials packaged in these granules is not clear.

The actual source of fixed negative charge within the granule

TABLE 6.2
Molecular Content of Mast Cells and Mast Cell Granules*

Molecule	pg/cell	pg/granule	Granule molar concentration
Histamine	10–40[1,2]	$10–40 \times 10^{-3}$	0.3–1.2
5-hydroxy tryptamine	0.7–1.0[3,4]	$0.7–1.0 \times 10^{-3}$	$1.3–1.9 \times 10^{-2}$
Heparin	40–100[5,6,7]	$40–100 \times 10^{-3}$	$1.8–4.5 \times 10^{-4}$ (0.6–1.7 equiv./L)
Protein	80–200[8,9]	$80–200 \times 10^{-3}$	
a) Rat mast cell chymase	52–130[10]	$52–130 \times 10^{-3}$	$6.9–17.3 \times 10^{-3}$
Multi valent metal cations			
$=Zn^{2+}$	(4×10^{-15} equiv./cell)[11]		(1.3×10^{-2} equiv./L)

* Granule contents were calculated by taking a value of 1000 granules/cell and granule volume of 0.3 $(\mu m)^3$ (Helander and Bloom, 1974).

1 Austen and Humphrey, 1963
2 Helander and Bloom, 1974
3 Moran *et al.*, 1962
4 Glick *et al.*, 1967
5 Fillion *et al.*, 1970
6 Slorach, 1971
7 Gustafsson and Enerbäck, 1978
8 Lagunoff *et al.*, 1964
9 Kawiak *et al.*, 1971
10 Yurt and Austen, 1977
11 Uvnäs *et al.*, 1975

with the potential to bind biogenic amines is also not clear. The binding of histamine to heparin has been shown to occur qualitatively at low ionic strengths, by empirical analysis of an equilibrium dialysis technique (Kobayashi, 1962) and by changes in induced Cotton effects of metachromatic dyes and ultraviolet optical rotatory dispersion (Stone, 1969). At physiological ionic strengths of NaCl (or at 0.003 M divalent ion) the binding of histamine to heparin is reduced to very low values (Kobayashi, 1962). The relationship of these observations to histamine binding to heparin *in vivo* is not known, as the ionic composition of the granule has not been studied.

Uvnäs *et al.* (1970) have suggested, on the basis of 1) the pH-dependence of histamine uptake by the granule in sodium chloride-free solution and 2) titration of dissociated, solubilized granules in 1M KCl and of commercial heparin alone (the titration behaviour of the granule protein was not studied), that the protein part of the granule may be essential for granule storage of histamine (for review see Uvnäs, 1977). It has been shown that histamine binding (5-hydroxytryptamine was also studied) to the granules at low ionic strengths is virtually abolished at pH <4, and it was claimed that the ionized carboxyl groups of the protein were essential for histamine binding. Granule uptake measurements of Na^+ or histamine at pH 7 yield a value of granule charge concentration near 1 μequiv/mg granule, which is approximately equal to granule histamine concentration in normal mast cells. An apparent problem in these fitrations focuses on whether there is partial dissociation of the granule with increasing ionic strength which occurs in Na^+ or histamine titration of the granule. The conclusion to be drawn from these studies is that carboxyl groups appear to be important in binding of histamine to the granule although the source of the carboxyl groups (i.e. either from heparin or protein) is not exactly defined.

On the basis of the amino acid composition of mast cell chymase (Everitt and Neurath, 1979) the maximum carboxylate charge concentration of the protein in the granule is calculated to be 0.2–0.6 equiv/L (assuming both aspartic and glutamic acids are totally in the acid form) (Table 6.2), this protein charge concentration could not totally account for histamine binding. At present, there are hardly any reasons why a protein/histamine complex may be singularly invoked to quantitatively explain histamine–granule interaction. In view of the high charge equivalence of heparin within the granule (Table 6.2), electroneutral complexes between histamine and heparin could be envisaged to account for major histamine fixation within the granule. The sulphate and carboxyl groups of heparin also appear to be involved in the formation of the insoluble complex with mast cell protein. The integrity of the histamine–protein–heparin granule complex would require a low ionic strength and/or low pH aqueous environment within the granule. (See also experiments on model granule systems Section 6.3). Furthermore, there is no support for the hypothesis of Padawer (1974) that the complex provides an 'ion' exchange vehicle for tissue fluid by removing toxic ions.

Although the mast cell granules are thought to be of a secretory nature, a secretory activity has not been demonstrated under normal *in vivo* conditions. Evidence for a secretory function of the mast cell comes from anaphylactic shock in dog (Jaques and Waters, 1941) in which heparin and histamine are released from liver mast cells into the circulation. This type of process has not been demonstrated elsewhere.

It is well established, however, that isolated mast cells may be triggered to release their granules by reaction of antigen with mast cell bound IgE, immunoglobulins or pharmacological agents. The mechanism of granule release (and its contents) from the mast cell has been subject to extensive study but appears to be essentially the same for rat mast cells for these various stimuli. A model for sequential exocytosis of mast cell granules as proposed by Uvnäs (1977) is shown in Figure 6.1. Mast cell granules exist within the cell coated by a perigranular membrane. The perigranular membrane may be partially removed by sonication of rat mast cells in sucrose (Anderson *et al.*, 1974). Membrane-free granules may be obtained by release from water-lysed mast cells, mast cells treated with compound 48/80 (condensation product of p-methoxyphenethylmethylamine with formaldehyde) (Anderson *et al.*, 1974) and by IgE antigen *in vitro* (Anderson *et al.*, 1973). It is to be noted that species differences apparently exist in the sensitivity of mast cells to degranulating agents, e.g. guinea pig mast cell is not sensitive to compound 48/80 and other agents (Uvnäs, 1977).

The first event seen in electron-microscope studies of the degranulation process (summarized by Uvnäs, 1977) is the positioning of peripheral granules to the cell membrane, with the formation of small foci of fusion between the granule membrane and the cell plasma membrane. The next morphological events to be observed are swelling of these peripheral granules, a reduction in their electron density and halo formation around them due to the loosening of the perigranular membrane. The cell membrane then ruptures and the granules, free of their perigranular membrane, are expelled into the extracellular fluid. High molecular weight [^{35}S]-heparin labelled *in vivo* or *in vitro* released from purified rat mast cells has been shown to be granule-associated and not released from the granule (Yurt *et al.*, 1977b). The heparin-protein granule complex appears to remain intact on release into the extracellular fluid, but the histamine is rapidly exchanged by Na^+. This exchange does not occur if the granule has retained its membrane (Anderson *et al.*, 1974). These membrane-free granules are able to retain their histamine only in deionized water. Solubilization of heparin and protein of the granule occurs when the granules have been repeatedly frozen and thawed (Bergqvist *et al.*, 1971) or at high KCl concentrations (1–2M) (Lagunoff *et al.*, 1964). Therefore, after the granule has been expelled from the cell and its perigranular membrane, it acts as a weak ion exchange resin. An anomalous feature of stimulated release of mast cell granule contents that is observed under certain conditions is the disproportionately greater release of histamine than of heparin (Lagunoff, 1972; Uvnäs, 1977). This has been interpreted as being due to incomplete release of the granule from the cell, in that the granules are exposed to the extracellular ionic milieu so the opportunity for histamine-exchange can occur, whereas the granule itself is retained within the domain of the cell or in cell cavities for potential reutilization of insoluble granule components such as heparin and protein.

Species differences in the solubility of the granules may determine the ultimate destiny for the heparin. For rat mast cells, the granules are insoluble in tissue fluids and are probably rapidly phagocytosed and digested by nearby macrophages (Riley, 1961; Higginbotham and Dougherty, 1956). On the other hand, mast cell granules from dog are soluble in tissue fluid and therefore will release heparin into blood under anaphylactic shock (Uvnäs, 1977).

Lindahl *et al.* (1979) have shown that [^{35}S]-labelled granules, when added to macrophage cultures *in vitro*, were taken up and the polysaccharide component converted into products that were largely released into the medium after extensive breakdown. The uptake of heparin (existing in a granule) and its subsequent degradation appears to be more efficient than heparin uptake alone stimulated by basic protein (Fabian *et al.*, 1978). Lindahl *et al.* (1979) have shown that macrophages were affected by granule uptake as noted by their spreading. It is of interest that the mast cell granule chymase appears inactive while in the granule form, due to its binding to heparin (Yurt and Austen, 1977), but this enzyme may be active in the macrophage. It is conceivable, then, that mast cells may have an important role in the regulation of some macrophage function.

A so far unrelated finding of interest is that heparin stimulates the accumulation of latent neutral proteinase and procollagenase in culture media of mouse bone implants (Sakamoto *et al.*, 1975; Vaes *et al.*, 1978). These enzymes are presumed to play a critical role in the resorption of bone or cartilage. Heparin not only increases the amount of enzyme released but may also stimulate collagenase activity (Sakamoto *et al.*, 1975; Werb and Burleigh, 1974). Although this heparin effect is not clearly understood, the fact that heparin may be taken up by the macrophage, in the light of evidence above, and stimulate secretion of hydrolytic enzymes would appear to be possible. On the other hand, Hauser and Vaes (1978) have shown that rabbit bone-marrow macrophages are able to degrade cartilage proteoglycans in culture without necessarily requiring heparin treatment.

These species differences are reminiscent of the fact that rabbit mast cells do not appear to contain heparin whereas mast cell heparin is abundant in rodents. This suggests that heparin may play a role in species specific initiation or regulation of chronic inflammation, through immunological-stimulated release from mast cells to uptake by macrophage with stimulated release of hydrolytic enzymes.

6.3 STORAGE AND SECRETORY GRANULES (NON MAST CELL)

There is now growing circumstantial evidence that sulphated glycosaminoglycans, including heparan sulphate, constitute and form an integral structural part of various granules in, perhaps, an analogous manner to the heparin-mast cell granule system.

There is also evidence that these glycosaminoglycans may play a role in the storage and release of hormones and biogenic amines in nervous tissue. Purified chromaffin granules isolated from bovine adrenal medulla contain small amounts of sulphated glycosaminoglycans (approximately 2% of the lipid-free dry weight) consisting of chondroitin 4- and 6-sulphate and a small portion of heparan sulphate (Margolis and Margolis, 1973) (see, however, Blaschke *et al.*, 1976). Heparan sulphate has been identified by use of electrophoretic and enzyme degradation techniques in various catecholamine subcellular particle fractions obtained from rat and cat heart and rat spleen. Catecholamine binding fractions were identified by the use of labelled noradrenalin injected into the animal prior to isolation of particles (Blashke *et al.*, 1976).

The amounts of glycosaminoglycans in the catecholamine binding compartments would appear to be too low to account for binding, by charge neutralization, of high quantities of amines. However, on the basis of studies of a model protamine–heparin complex for an ion exchange store for biogenic amines, Uvnäs and Åborg (1976) have suggested that glycosaminoglycan-containing protein granules may have a high affinity for amines, which are only dissociated at high ionic strength. The heparin concentration used in these models is relatively high; the initial mixing ratio approximated equal weight ratios of the macro-ions. Uptake studies of the protamine–heparin complex for either sodium, adrenaline or phenylethylamine up to $10 \mu g/ml$ of micro-ion all showed the same binding equivalence to saturatable sites on the complex. The binding showed marked pH dependence and at pH4 there was essentially no uptake, suggesting that the carboxyl groups of either heparin or protein may be involved in this binding. At higher micro-ion concentrations (0.01–0.5M) a second phase of uptake was observed. At these concentrations, dissociation of the complex was observed making available more sites for binding. At present, these results are difficult to interpret in terms of different conditions and binding behaviour associated with the micro-ion interaction with the complex, and the relationship of this system to the *in vivo* micro-ion environment.

Giannattasio and Zanini (1976) have suggested the presence of heparin, chondroitin 4-sulphate, chondroitin 6-sulphate and heparan sulphate in purified rat prolactin secretory granules from the rat pituitary gland. These organelles can be isolated as a pure fraction, and retain their structural organization after solubilization of their limiting membrane by mild detergent treatment. The sulphated glycosaminoglycans were labelled, when pituitary slices were incubated *in vitro* with [^{35}S]-sulphate, and subsequently identified by electrophoretic techniques. The concentration of sulphated glycosaminoglycans was relatively low and estimated to be 60 μg uronic acid/mg granule protein, 80% of which is prolactin. Studies on rat exocrine pancreas (Kronquist *et al.*, 1977) labelled *in vitro* with [^{35}S]-sulphate and subsequent subcellular fractionation showed that radioactivity was incorporated into the following fractions; zymogen granules 14%, zymogen granule membranes 3%; soluble zymogen granule contents 7%; microsomes 30%; mitochondria 14% and post microsomal supernatant 25%. Heparan sulphate was the major glycosaminoglycan present in all the subcellular fractions (identified by ion exchange chromatography, digestion with heparinase and by deaminative cleavage with nitrous acid). Determination of the level of glucuronic acid in the zymogen granule indicated that sulphated glycosaminoglycans were present at levels of approximately 0.5 nmoles/mg of zymogen granule protein.

6.3.1 Granule formation and function

It would appear that the concentrations of sulphated glycosaminoglycans are too low to invoke a charge neutralization mechanism with granule protein, with the possible exception of heparin-containing mast cell granules. However, the importance of these small quantities of heparan sulphate and other glycosaminoglycans has been postulated in the formation of zymogen granules in pancreatic cells. After synthesis of secretory proteins, they are transported from the rough endoplasmic reticulum to the Golgi region where they are concentrated and packed for storage in zymogen granules until their release by exocytosis. It is thought that the sulphated glycosaminoglycans synthesized in the Golgi complex (Berg and Young, 1971; Young, 1973; Reggio and Palade, 1978; Berg, 1978) are involved in the packaging of secretory proteins, at the Golgi, to form secretory granules (for review of these concepts see Jamieson and Palade, 1977). On discharge from the cell, the granules will dissociate at higher pH to allow enzyme solubilization in the pancreatic juice. Although not studied explicitly, the mechanism of granule formation in other systems including mast cells, may also occur by the same route.

Model experiments have shown that sulphated glycosaminoglycans may interact with cationic proteins to form granule-like aggregates. Reggio and Dagorn (1978) have studied the interaction between bovine chymotrypsinogen A and chondroitin sulphate or sulphated glycosaminoglycans isolated from guinea pig zymogen granules and found both preparations were able to induce aggregate formation of chymotrypsinogen in a similar fashion as measured by turbidimetric techniques. They found that precipitation was not affected by calcium and magnesium ions at concentrations known to cxist in thc guinca pig zymogen granules ($\sim$37 and $\sim$9 nmole/mg granule protein respectively Clementi and Meldolesi, 1975). Certainly, at low glycosaminoglycan concentrations they found marked precipitation (i.e. 1 μg/ml chondroitin sulphate and 1 mg/ml chymotrypsinogen) with the bulk of the glycosaminoglycan being incorporated into the complex. They suggest that this behaviour could account for the granule formation properties of these molecules *in vivo*. Of interest in the studies of Reggio and Dagorn (1978), is that the co-precipitation of chymotrypsinogen with chondroitin sulphate showed concentration dependence with respect to the glycosaminoglycan, with maximal precipitation occurring at low concentrations of glyco-

TABLE 6.3

Co-precipitation of Bovine Chymotrypsinogen (ChTg) with Chondroitin Sulphate (CS) and Sulphated Glucosaminoglycan (GAG-SO_4) from guinea pig zymogen granules

(From Reggio and Dagorn, 1978 with permission of the publishers.)

CS added (% of ChTg wt/wt)	ChTg precipitated (% of total)	GAG-$^{35}SO_4$ in the pellet (% of total counts)
0	<5	65.6
2	30.2	98.7
4	68.5	97.1
6	82.0	99.4
6.5	90.0	98.3
8	87.2	96.6
10	70.0	84.6
12	42.9	63.1
14	24.3	31.7

The mixture contained in 1 ml of 2 mM Tris maleate, pH 6.0: 1 mg of ChTg, 50 μl of GAG-$^{35}SO_4$ solution containing 1,500 cpm and various amounts of CS. When ChTg and ChTg+CS were omitted, GAG-SO_4 precipitation was 0.5 and 2%, respectively.

saminoglycan. Gradual inhibition of aggregation is observed at relatively higher concentrations of chondroitin sulphate with proportionately less of the polysaccharide being incorporated into the complex (Table 6.3). It is evident that at low concentrations, the sulphated glycosaminoglycan may modify protein–protein interactions more efficiently.

Similar data have come from experiments on the interaction of rat mast cell chymase and macromolecular heparin, which may result in the formation of insoluble complexes in 0.0075M Tris pH 7.4 (Yurt and Austen, 1977). The insoluble complex formation was studied by monitoring sedimentable material in sucrose density gradients and was shown to be strongly dependent on the concentration ratio of the reactants. Maximal insoluble complex formation occurred at a molar ratio of chymase:heparin of 40:1 (and was in accord for molar ratios of these materials found in mast cell granule Table 6.2). By increasing heparin concentration to molar ratios 8–16:1, no insoluble complex was formed.

Certainly, a striking characteristic of storage granules considered is the combination of glycosaminoglycan (or proteoglycan)/proteolytic enzyme (in an inactive form). This is analogous to proposed function of lyosomal glycosaminoglycans inactivating lysosomal enzymes. Although there are few documented data on the presence of heparin-like polysaccharides in lysosomes, studies have demonstrated that most lysosomal enzymes interact with glycosaminoglycans in a pH-dependent, reversible, electrostatic manner, which results in enzyme inhibition (Avila and Convit, 1975, 1976). The strength of interaction and enzyme inhibition was much greater at pH 4.0 than at pH 5.0 and the order of strength of inhibition at pH 4 was heparin >chondroitin 4-sulphate ≃ chondroitin 6-sulphate > dermatan sulphate (Avila, 1978). It has been suggested that these enzymes are in a protected and inactive form while in the acid medium of primary lysosomes. When phagocytosis of microorganisms or other cells by leucocytes occurs, a sequence of events follow which lead to the formation of a phagocytic vacuole (phagosome), the fusion of a primary lysosome with a phagosome which will result in dilution of the intralysosomal acid fluid and therefore an increase in pH. This, in turn, could lead to a dissociation of the complexes and a release of active enzyme, which will ultimately digest the engulfed material.

6.4 CELL MEMBRANES

6.4.1 Polysaccharide localization

The ubiquitous presence of heparan sulphate in various tissues of the body undoubtedly reflects the now well established finding of Kraemer (1971a) that heparan sulphate is probably a general constituent of cell membranes. Kraemer (1971a) demonstrated that when exponentially growing Chinese hamster cells were supplied with radioactive glucosamine and inorganic sulphate for a 21 h growth period (1 generation), radioactivity from both sources was found in a variety of complex carbohydrates, including heparan sulphate, in three cellular fractions; as a cell surface component removable with trypsin treatment; as a free, directly acid-soluble component of the 'cell sap' and as part of the residual acid precipitate that became acid soluble following papain digestion. Almost 60% of the radioactivity, identified as heparan sulphate, appeared to be free in the cytoplasm and directly acid soluble, while the remainder appeared to be mainly membrane-associated, requiring proteolysis for its release.

Furthermore, a number of investigators at this time had shown that heparan sulphate was synthesized by a variety of cells. Kraemer (1971b) demonstrated the ability of six established mammalian cell lines and a diploid mouse embryo lung strain to make heparan sulphate. Dietrich and De Oca (1970) detected heparan sulphate biosynthesis in two established cell lines, as well as in cells grown *in vitro* from disaggregated mouse and rat embryos. Heparan sulphate biosynthesis by cells *in vitro* also has been demonstrated by Dorfman and Ho (1970) in glial tumor cells, by Buonassisi (1973) in endothelial cells, by Buonassisi and Ozzello (1973) in synovial sarcoma cells, and by Satoh *et al.* (1974) in B-16 mouse melanoma cells. In view of the possible changes that these cells may undergo in adapting to *in vitro* conditions, Conrad and Hart (1975) have shown that the ability of three defined fibroblast populations to make heparan sulphate was not acquired as an *in vitro* response of the cells to their particular inoculation conditions, (as studied 10 h after inoculation) and concluded that many cell types *in vivo* may normally make heparan sulphate.

Further evidence for the localization of heparan sulphate as a bona fide cell membrane component has come from the release of heparan sulphate when cells interact with exogenous heparin-like polysaccharides. Kjellén *et al.* (1977) have shown saturatable, temperature-dependent, reversible binding of exogenous heparin and heparan sulphate to rat liver cells, which suggested that endogenous cell-surface heparan sulphate represents material bound to heparin-like polysaccharide exchangeable receptor sites. The specificity of binding was demonstrated by adding various glycosaminoglycans to cultures containing cells prelabelled with inorganic [^{35}S]-sulphate; only heparin and certain heparan sulphates but none of the other polysaccharides tested, including hyaluronate, chondroitin sulphates, dermatan sulphate, and some preparations of low-sulphated heparan sulphate were able to displace the endogenous polysaccharide (Kjellén *et al.*, 1977; Kraemer cited in Lindahl and Höök, 1978). Kraemer demonstrated that heparan sulphate of the cell surface of cultured Chinese hamster cells is released when prelabelled [^{3}H]-glucosamine cells were incubated with solutions containing heparin. One third to one half of the cell surface heparan sulphate (i.e. trypsin-releasable material) is released by heparin in 30 minutes at 37°C. The distribution of released material on Agarose 0.5M showed that the released heparan sulphate existed as two distinct populations of high and low molecular weight material. The distribution was not affected by trypsin treatment. These observations point to the possibility that tryptic release of heparan sulphate from the cells may be due to the attack of receptor sites which bind cell surface heparan sulphate chains, rather than the peptide moiety of the bound proteoglycan. Further work is required to understand the mechanism of tryptic release however.

Another approach employed in the establishment of heparan sulphate localization on the cell membrane has come from the use of specific enzymes known to degrade these glycosaminoglycans. Buonassisi and Root (1975) demonstrated with cultures of endothelial cells, prelabelled with [^{35}S]-sulphate, when exposed to a purified preparation from induced *Flavobacterium heparinum* containing heparinase and heparitinase activities, radioactivity accumulated in the supernatant and medium. Protamine sulphate, which binds strongly to heparin-like polysaccharides, inhibited enzyme degradation. They suggested that the chains of heparan sulphate that are accessible to the action of the enzyme are present at the surface of endothelial cells. Wasteson *et al.* (1977) have used heparan sulphate endoglycosidase to degrade prelabelled heparan sulphate of human glial cells. For rat liver cells, Oldberg *et al.* (1977) found that heparitinase treatment released less than half of the trypsin-sensitive [^{35}S]-labelled cell surface polysaccharide. The same result was obtained for purified plasma membranes which suggests a restricted susceptibility or accessibility of cell surface heparan sulphate to the enzyme.

In view of the ambiguities raised by the use of enzyme-treated cells to distinguish specific cell localized glycosaminoglycans, as the enzymes may disrupt or damage some cells leading to the release of intracellular material, a number of investigators have established the occurrence of heparan sulphate in purified plasma membranes (Yamamoto and Terayama, 1973; Funakoshi *et al.*, 1974; Akasaki *et al.*, 1975; Nakada *et al.*, 1975; Oldberg *et al.*, 1977). Recent investigations by Oldberg *et al.* (1979) have demonstrated that solubilization of heparan sulphate from rat liver plasma membranes could not be obtained by incubation of the membranes with the cation chelator EDTA, suggesting that calcium is not involved in cell-surface heparan sulphate binding. Extraction of the membrane fraction with 2M NaCl or treatment with heparin released 65% of the membrane-associated heparan sulphate. The residual heparan sulphate could be removed with detergent.

6.4.2 Metabolic activity of cellular heparan sulphate

A preponderance of heparan sulphate on the cell surface is not always evident and it may vary widely with cell type and with some fundamental aspects of cell behaviour (see following section). In discussing the general features of heparan sulphate in different environs of the cell, its existence in four regions should be considered. The presence of heparan sulphate at the cell surface, also referred to as the 'ectocellular' or the pericellular region has been discussed in the previous section. Heparan sulphate also exists intracellularly, which includes 'cell sap' and 'residual precipitate fraction' (Kraemer, 1971a; Kraemer, 1977) and may also be secreted into the extracellular medium (Roblin *et al.*, 1975). Heparan sulphate has also been identified in an 'undercellular' pool (also referred to as surface-attached material which is associated with the attachment of cells to the container in which they are grown in culture) by [^{35}S]-incorporation into heparan sulphate isolated from that region (Kresse *et al.*, 1975a; Rollins and Culp, 1979).

It can be stated at the outset that the metabolic relationship of the heparan sulphates found in these four regions is by no means clear. Only fragmentary and isolated evidence has been published concerning the chemical and physical properties of the heparan sulphates found in these regions of the cell.

One feature that has emerged from studies on heparan sulphate metabolism in cell cultures is that the metabolic activity of heparan sulphate on the cell surface and in the extracellular region is high. In the initial studies of Kraemer (1971a), the metabolic relationships of heparan sulphate in three areas were studied, i.e. trypsin removable cell surface component, cell 'sap' and a residual precipitate by a tandem-labelling method in which the cells were grown in the presence of [^{3}H]-glucosamine and then given a final 4 h pulse with [^{14}C]-glucosamine. In this case, heparan sulphate isolated from the surface and cell-sap and fractionated on DEAE-cellulose had distinctly different ^{3}H/^{14}C ratios (1.61 and 6.26 respectively). This suggested that the rate of cell surface heparan sulphate synthesis is relatively high and that it did not represent 'leaching out' of cell-sap material during the trypsin treatment.

The relatively high metabolic activity of cell surface heparan sulphate in rat liver cells has also been reported in more recent studies by Oldberg *et al.* (1977) where incorporation of label into this molecule reaches a steady state after 10 h. Since it had been established that heparan sulphate is secreted into the extracellular space (Robin *et al.*, 1975) studies on the rates of incorporation of [^{35}S]-sulphate into heparan sulphate of human embryonic lung fibroblasts demonstrated that the highest rates were found for the medium and trypsin digests of cells, whereas low rates of incorporation were found for trichloroacetic acid extracts of the cells and the cell residue (Sjöberg and Fransson, 1977).

In studies of cultured bovine arterial cells Kresse *et al.* (1975b) found that the rates of incorporation of [^{35}S]-sulphate into heparan sulphate were of the order extracellular $>$ pericellular $>$ intracellular. The half-life of intracellular heparan sulphate was 7–8 h and was mainly in the process of degradation as indicated by its low molecular weight. The pericellular pool of heparan sulphate gave a half-life of 24 h. The distribution of heparan sulphate synthesized was recorded as 49% in the degradative pool, 26% pericellular pool and 25% extracellular. Kresse *et al.* (1975b) also showed that when arterial proteoglycans are added to the culture medium they are integrated into the pool of cell membrane-associated (trypsin-removable) glycosaminoglycans by a saturatable process, which depends on time and temperature. The glycosaminoglycans are then transported to the interior of the cell for degradation in lysosomes. This process of uptake, called pinocytosis, is optimized when the proteoglycan has an intact protein moiety and presumably undegraded carbohydrate chains of the proteoglycan. Only a small part of the cell surface pool represents material destined for immediate pinocytosis, whereas most represents newly synthesized material derived from the extracellular space.

Although cell surface heparan sulphate may exist as a high molecular weight proteoglycan form (Section 4.3.2), the molecular size of heparan sulphate in other regions of the cell remains to be clearly defined.

6.4.3. Changes of heparan sulphate associated with particular cellular events

The widespread occurrence of heparan sulphate in cellular environments indicates that it may have important biological functions associated with cell behaviour. In particular, the localization of heparan sulphate in the plasma membrane may suggest its involvement in cell–cell interaction and cell–extracellular matrix interactions. The possible role of heparan sulphate in cell behaviour has been studied in cell cultures. In such systems it is envisaged that it may influence cell proliferation, cell adhesion, cell aggregation and accessibility of cell surface receptors to external or internal factors. A large body of data now exists which demonstrates an empirical relationship of the relative amounts of heparan sulphate and other glycosaminoglycans of the cell surface with some fundamental aspects of cell behaviour, although the etiology of these changes and their relationship to cellular events is not clearly understood. At this stage, there is no compelling experimental evidence to define the biological function of cellular heparan sulphate.

Initial studies in this direction were performed by Kraemer and Tobey (1972) who showed pronounced premitotic loss of surface heparan sulphate in Chinese hamster cells undergoing cell division. The loss of heparan sulphate corresponds to the release of the intact molecule to the medium. A more general, contemporary view emerging from studies on cell growth and DNA-virus and RNA-virus tumor transformed cells, where in both cases a decrease in heparan sulphate on the cell surface occurs, is that the heparan sulphate may, in part, play a role in negative control of cell growth and favour homeostasis (Chiarugi and Vannucchi, 1976; Chiarugi, 1976; Augusti-Tocco and Chiarugi, 1976). There is, however, conflicting evidence on this score as other investigations have shown that these cellular events are not concomitantly accompanied by such changes in heparan sulphate (Cohn *et al.*, 1976; Glimelius *et al.*, 1978b; Chiarugi and Dietrich, 1979).

There is evidence to suggest that the quantitative changes of heparan sulphate associated with these cellular events are also accompanied by changes in the chemical composition of this polysaccharide (Underhill and Keller, 1976, 1977; Keller *et al.*, 1978). At least two types of cell surface heparan sulphate have been identified in normal mouse embryo cells, designated I and II. These two forms have been distinguished by both ion-exchange chromatography on DEAE-cellulose and electrophoresis at pH 1, which probably reflects differences in the degree of sulphation. The molecular weights of these heparan sulphates appear identical as judged by gel chromatography. Type II heparan sulphate has similar properties (on DEAE-columns and electrophoretic behaviour) to that heparan sulphate produced by several mouse cell lines that exhibit contact inhibition of growth. Type I heparan sulphate has similar properties to the predominant heparan sulphate derived from either DNA or RNA virus-transformed cell lines which lack control (Underhill and Keller, 1976, 1977). Similar chemical changes have been monitored for both medium and intracellular heparan sulphate with viral transformation (Johnston *et al.*, 1979).

It is of interest that, whereas virally transformed cells may induce changes in heparan sulphate composition, exogenous heparin may inhibit virus infectivity (Choi *et al.*, 1978 and earlier references). Once virus absorption takes place, heparin has no effect.

6.5 CELL MEMBRANE MEDIATED INTERACTIONS

6.5.1 Binding of exogenous heparin-like polysaccharides

Initial studies of the binding of exogenous heparin to rat liver cells (Kjellén *et al.*, 1977) and Chinese hamster cells (Kraemer, 1977) showed that the binding of heparin was saturatable, temperature dependent, reversible and displaced cell surface heparan sulphate. These studies thus demonstrated that endogenous heparan sulphate or exogenous heparin may bind to receptor sites on the cell surface membrane. Further confirmation of these studies has come from work on the binding of [^{3}H]-heparin to the surface of cultured human endothelial cells (Glimelius *et al.*, 1978a) (see also Hiebert and Jaques, 1976ab) which has also been shown to be time dependent, saturatable, reversible trypsin sensitive and unaffected by the presence of excess amounts of other glycosaminoglycans such as chondroitin sulphate, dermatan sulphate, hyaluronic acid or heparan sulphate. Both high affinity heparin and low-affinity heparin for antithrombin III (Section 7.2.1.2) showed similar binding behaviour (Glimelius *et al.*, 1978a). The binding capacity corresponds to about 10^6 molecules of heparin per cell. The binding of unfractionated [^{3}H]-heparin with platelets has also been demonstrated (Shanberg *et al.*, 1976). The platelets were found to take up only free heparin but not heparin complexed with antithrombin III. The binding is saturatable and reversible.

The binding of exogenous heparin or other heparin-like polysaccharides or the presence of endogenous heparin-like polysaccharides on the cell surface may then confer on this region a number of important properties 1) antithrombotic properties (Section 7.1) 2) localization of exogenous enzyme activity such as the binding of lipoprotein lipase which may then act on lipoprotein substrates (Olivecrona *et al.*, 1977) (Section 7.6) 3) affect cell membrane activity such as membrane bound adenylate cyclase (Section 6.5.3) and 4) mediate cell surface–extracellular matrix interactions (Section 6.5.2).

6.5.2 Cell membrane–extracellular matrix interaction

In terms of membrane structure, the fluid mosaic model for the structure of the plasma membrane as postulated by Singer and

Nicolson (1972) involves the existence of integral proteins that, within limits, are free to diffuse laterally in the plane of the membrane formed by a fluid phospholipid bilayer and peripheral membrane proteins bound to the surface of the bilayer. The lateral mobility of membrane components may have an important role in the transmission of signals through the membrane together with cooperative interactions on the cell membrane surface. The organization and distribution of heparan sulphate and other cell-surface glycosaminoglycans on the cell surface is not known, but it is conceivable that they are bound to protein receptors within the plasma membrane structure and therefore may have lateral mobility along the membrane surface. It is likely that heparan sulphate, being a highly charged polyanion, will contribute relatively high anionic charge density to its local area and therefore will potentiate the binding of proteins to the cell surface. Absolute quantitative estimates of the surface concentration of heparan sulphate have not been made, but it is a relatively minor component, as compared to glycoprotein and other macromolecules on the cell surface. A value of about 10^6 molecules of exogenous-bound heparin per cell to the surface of cultured human endothelial cells has been reported (Glimelius *et al.*, 1978a). While heparan sulphate has been shown to participate in many diverse interactions with proteins, whether these are important on the cell surface remains to be investigated. In any case, as it appears to be a general cell surface component it obviously is a potential candidate for participation in a range of cell membrane mediated interactions including cell migration, cell shape, cell adherence, cell–cell contact and cell–extracellular matrix interactions.

There is now accumulating evidence that heparin-like polysaccharides on the cell surface may mediate and facilitate the interaction of the cell surface with extracellular matrix proteins. Öbrink (1973) has studied the binding of various connective tissue glycosaminoglycans, including heparin and heparan sulphate to monomeric lathyritic collagen by light scattering, turbidimetric and precipitation techniques. The binding appeared to be essentially electrostatic, under physiological conditions, as the interaction is abolished at high ionic strength. Qualitative estimates of binding could only be made, but these suggested that the heparin-like polysaccharides may bind strongly to collagen. Solution studies have also shown that the connective tissue polysaccharides may affect collagen fibril formation although their degree of effectiveness has been subject to conflicting reports (for review see Comper and Laurent, 1978). It is conceivable that heparan sulphate and other cell surface glycosaminoglycans may alter the rate of collagen fibrillogenesis occurring at or near the cell surface area.

Another extracellular matrix protein known to be associated with the membrane surface of various cells is fibronectin. This is a glycoprotein, or a group of glycoproteins, variously known as cell surface protein, fibroblast surface antigen and LETS (large external transformation sensitive) protein (Bornstein *et al.*, 1978). The protein has been found on the surface of various cells including fibroblasts, smooth muscle cells, glial cells and both epithelial and endothelial cells (for review, see Bornstein *et al.*, 1978). In addition, a protein similar in molecular weight and which is immunologically related to cell surface fibronectin, is a soluble plasma component termed cold insoluble globulin. Although the direct interaction of heparin-like polysaccharides with cell surface fibronectin has not been studied a number of investigations point to potential importance of this reaction in affecting cell behaviour. The interaction of heparin with cold insoluble globulin has been shown to form soluble complexes at room temperature and insoluble complexes in plasma at low temperatures (Stathakis and Mosesson, 1977). As fibronectin demonstrates haemoagglutinin activity (Yamada *et al.*, 1975) it is also of interest that heparin-like polysaccharides show antihaemoagglutinin activity with embryonic chick muscle tendon lectins (Kobiler and Barondes, 1979). Fibronectin is reduced or absent from the surfaces of many transformed and malignant cells of mesenchymal origin (Hynes, 1976; Vaheri *et al.*, 1976). Its metabolic control may be linked to cell surface heparan sulphate which is also known to decrease quantitatively under similar conditions (Section 6.4.3). The close association of fibronectin and heparan sulphate has also been suggested in terms of their enrichment in the adhesion sites of reattaching cells to artificial substrates (Rollins and Culp, 1979). These adhesion sites represent extracellular matrix material synthesized by the attaching cell to form Ca^{2+}-requiring footpads by which the cell adheres to substrate. The concept that heparan sulphate may mediate interaction with fibronectin and collagen in these initial adhesion sites has also been proposed (Jilek and Hörmann, 1979).

6.5.3 Adenylate cyclase interaction

One feature of the binding of heparin-like polysaccharides to plasma membranes, in modifying membrane and cell behaviour, has recently come from studies on plasma membrane bound adenylate cyclase. This enzyme has important functions associated with the control of cell proliferation, differentiation and malignant transformation (Kelley *et al.*, 1977 and references therein). The fact that heparin-like polysaccharides may affect adenylate cyclase activity may be important in controlling cell behaviour.

Heparin has been shown to be a sensitive inhibitor ($I_{50}=2$ μg/ml) of luteinizing hormone- and human chorionic gonadotropin-stimulated adenylate cyclase activity in rat ovarian plasma membranes (Salomen and Amsterdam, 1977; Salomen *et al.*, 1978). The heparin had a weaker effect on hormone-independent activity. The inhibition effect of heparin appears to be a general phenomenon as inhibition, to varying extents, was also apparent when this enzyme was stimulated by follicle-stimulating hormone or prostaglandin E_2. Inhibition by heparin was also registered for glucagon sensitive-rat hepatic adenylate cyclase and the prostaglandin E-sensitive enzyme from rat ileum and human platelets. In contrast, heparin stimulated the dopamine sensitive adenylate cyclase from rat caudate nucleus (Amsterdam *et al.*, 1978).

In comparative studies, Salomen and Amsterdam (1977) have shown that heparin, heparan sulphate, dermatan sulphate and dextran sulphate are potent inhibitors of ovarian adenylate cyclase in which heparin was most effective (heparin $I_{50}=2$ μg/ml, dextran sulphate $I_{50}=6$ μg/ml; dermatan sulphate $I_{50}=16$ μg/ml, heparan sulphate $I_{50}=20$ μg/ml). Chondroitin sulphate and hyaluronate were without effect (Follicular fluid itself obtained from medium or large bovine ovarian follicles inhibited the ovarian adenylate cyclase ($I_{50} = 3$ mg follicular fluid protein/ml)). The mechanism of inhibition by heparin is not clear, but has been attributed to its effect either on the binding of the hormone to membrane receptor sites or its interference with enzyme receptor coupling by restricting the membrane mobility of these components or by affecting their conformation (Salomen *et al.*, 1978). Similar conclusions were made by Kelley *et al.* (1977) in removing the glycosaminoglycan components (unidentified) of embryonic cultured human diploid fibroblasts by digestion with testicular hyaluronidase (which removes 40% of the glucosamine-containing material of the membrane). The removal of cell surface glycosaminoglycan components by this method was associated with stimulation of norepinephrine-stimulated adenylate cyclase as compared to its activity in intact membranes; as a result, cell behaviour changed significantly.

A recent study of Gebauer *et al.* (1978) has shown that when rat ovarian slices were incubated with culture medium containing [^{35}S]-sulphate, they produced a number of radiolabelled glycosaminoglycans, including heparin-like substances as characterized by nitrous acid sensitivity and electrophoretic behaviour. The concept that these heparin-like polysaccharides may participate in ovulation-related events in the ovary, through their interaction with cell membrane adenylate cyclase remains to be investigated.

6.6 CELL NUCLEUS

There is circumstantial evidence to suggest the presence of glycosaminoglycans, including small quantities of heparin-like polysaccharides, in cell nucleii. Early work by Pritchard *et al.* (1971) showed that glucosamine-containing glycosaminoglycans copurified with DNA preparations obtained from several different tissues of animals. The DNA was isolated by an aqueous polymer–two phase procedure and subjected to CsCl equilibrium density gradient centrifugation. More recent studies have focused attention on nucleii isolation and isolation of glycosaminoglycans by more conventional means. These studies include isotope incorporation into glycosaminoglycans isolated from nucleii of cultured B16 mouse melanoma cells (Bhavanandan and Davidson, 1975), Hela S_3 cells (Stein *et al.*, 1975) and quantitative chemical analysis of glycosaminoglycans from rat brain nucleii (Margolis *et al.*, 1976) and rat liver nucleii (Furukawa and Terayama, 1977). The localization of [^{35}S]-labelled glycosaminoglycans within cell nucleii of skin fibroblasts cultured from patients with Sanfilippo's disease type B and Hurler'sisease by high resolution autoradiography has been reported by Fromme *et al.*, (1976).

The total quantities of glycosaminoglycans in cell nucleii appear quite small. Margolis *et al.* (1976) have reported a value of 0.142 μmol hexosamine/100 mg protein for rat brain nucleii and Furukawa and Terayama (1977) have given a value of 0.2–0.3 μg hexuronic acid per mg DNA for rat liver nucleii. Furthermore, the quantities of heparin-like polysaccharides within the nucleii glycosaminoglycan population are low. Heparin has yet to be identified within cell nucleii and only trace quantities of heparan sulphate have been found (Bhavanandan and Davidson, 1975; Margolis *et al.*, 1976; Furukawa and Terayama, 1977). It would also appear that the glycosaminoglycans in general, are minor nucleii components compared to glycoproteins also studied by these workers.

In view of the small quantities of glycosaminoglycans that have been found associated with nucleii, great care must be taken in purifying nucleii free of subcellular contaminants, including plasma membranes which are known to contain heparan sulphate (Section 6.4). Furukawa and Terayama (1977) have used several approaches to establish that the glycosaminoglycans associated with nucleii were bona fide nucleii components rather than being derived from contaminants. Apart from microscopic examination of purified nucleii, they performed analysis of enzyme markers for plasma membranes and mixed labelled glycosaminoglycans with crude nucleii fraction prior to purification of nucleii. The actual site of synthesis or source of nuclear glycosaminoglycans is not known. The conclusion to be drawn at the present time is that nuclear sites for glycosaminoglycan localization, particularly the heparin-like polysaccharides, is a tentative proposition.

6.6.1 Template activity

The function of nuclear glycosaminoglycans is unknown, although significant investigations in this area have occurred over the last few years. Studies with model experiments have suggested that exogenous polyanions (particularly heparin) may affect nuclear enzyme activity and control template activity in chromatin.

Heparin has been shown to stimulate DNA and RNA synthesis *in vitro*. This has been interpreted as being due to the direct interaction of heparin with histones and non-histonal proteins of nucleii, resulting in derestriction of the DNA template and stimulation of DNA and RNA synthesis. In particular, heparin has been shown to affect chromatin morphology (noticeably inducing swelling) (Kraemer and Coffey, 1970a; Arnold *et al.*, 1972; Miller *et al.*, 1972; Cook and Aikawa, 1973), destabilization of chromatin to thermal denaturation (Ansevin *et al.*, 1975; Kinoshita, 1976; Smith and Cook, 1977), chromatin susceptibility to DNase (Saiga and Kinoshita, 1976; Janekedivi, 1978) and increased template activity of chromatin (as measured by the increased incorporation of labelled deoxyribonucleosides) for both exogenous and endogenous DNA and RNA polymerases (Chambon *et al.*, 1968ab; Miller *et al.*, 1972; Kraemer and Coffey, 1970b; Berlowitz *et al.*, 1972; de Pomerai *et al.*, 1974; Warnick and Lazarus, 1975; Kitzis *et al.*, 1976). A direct relationship between heparin-induced changes on chromatin structure as described above and concomitant changes in template activity has yet to be established (see for example Smith and Cook, 1977).

The effects of heparin on template activity appear to be concentration dependent. A general view emerges that at relatively low heparin concentrations (see below) preferential binding of heparin to histones and non histonal proteins takes place resulting in derestriction of template activity. At relatively higher concentrations of heparin, histonal binding becomes saturated, and an inhibitory effect is then observed as excess heparin interacts with both DNA and RNA polymerases (Warnick and Lazarus, 1975; Seki and Oda, 1977; Coupar and Chesterton, 1977). The scale of heparin concentrations required for these bimodal effects appears to be different for endogenous and exogenous enzyme activities, where greater heparin concentrations are required to affect the latter (Smith and Cook, 1977).

It is well documented that heparin may act as a potent inhibitor of the free form of the DNA and RNA polymerases and certain nonspecific polymerase-DNA complexes (Walter *et al.*, 1967; Zillig *et al.*, 1971; Schäfer *et al.*, 1973; Ferencz and Siefart, 1975; Schaffrath *et al.*, 1976; Fabry *et al.*, 1976; Coupar and Chesterton, 1977; Brennessel *et al.*, 1978; Di Cioccio and Srivastava, 1978) and that this inhibition is concentration dependent. (The binding of heparin to these enzymes has been used for their purification on heparin-linked Sepharose columns.) In certain instances heparin is, however, without effect on transcribing enzymes where the degree of specificity between the enzyme and DNA is high; as in the formation of the complex *in vivo* or after the onset of initiation *in vitro* as shown for both RNA polymerases (Siefart *et al.*, 1973; Schäfer *et al.*, 1973; Ferencz and Siefart, 1975; Coupar and Chesterton, 1977) and DNA polymerases (Schaffrath *et al.*, 1976; see however Di Cioccio and Srivastava, 1978).

In studies on heparin–histone interaction, it appears that heparin may effectively compete with DNA and other proteins in their interaction with all histones (H1, H2A, H2B, H3 and H4) and possibly with non-histone proteins in chromatin. Early studies by Kent *et al.* (1958) showed that heparin may form strong complexes with isolated histones. More contemporary studies have shown the activity of heparin to form stable complexes with isolated histones appears to be different to its interaction with histonal proteins in the intact chromatin (Kitzis *et al.*, 1976; Taylor and Cook, 1977; Hildebrand *et al.*, 1977) thus suggesting a specific organization of these proteins in the chromatin structure. The specificity of heparin–histone interaction is not clearly understood, however. The treatment of chromatin with heparin appears to remove preferentially lysine-rich histones (H1) (Hildebrand *et al.*, 1977). The removal of lysine-rich histones by other techniques is known to stimulate endogenous DNA polymerase activity (Janakidevi, 1978). There is some evidence to suggest that heparin may also inhibit histone phosphorylation (Mäenpää, 1977) and the phosphorylation of histone H1 results in its decreased susceptibility to interact with heparin (Hildebrand and Tobey, 1975; Walters and

Hildebrand, 1975).

In general, the interaction of heparin with nuclear components appears to be due primarily to its charge density. More highly charged polysaccharides (normally charged by sulphation), such as dextran sulphate, have been shown to exert greater effects on template activity and polymerase activity than heparin. Other glycosaminoglycans of lower charge density, such as chondroitin sulphate and hyaluronic acid, have little or no effect on these interactions (Arnold *et al.*, 1972; Warnick and Lazarus, 1975; Schaffrath *et al.*, 1976; Brennessel *et al.*, 1978). There is no compelling evidence that these effects are the result of increasing polysaccharide sulphation *per se* as compared to increasing polysaccharide charge density and therefore decreasing the mean intercharge distance in a particular manner (see Section 5.2.1 for general comments on the effect of polysaccharide charge type on macromolecular electrostatic interactions).

While heparin may produce significant effects on nuclear processes *in vitro* it is difficult to make quantitative comparisons of the observed effects at this stage. Certainly, the concentrations of heparin used far exceed the reported nucleii glycosaminoglycan levels and further, heparin has yet to be identified as a nuclear component. In view of the multiple interactions that heparin can potentially undertake with nuclear proteins, the effects observed within a system will be in part a function of preferential binding of heparin to a particular substrate and the concentrations of the other interacting components. For enzyme studies, in particular, there is little quantitative significance in the inhibitory parameters reported in view of the above considerations, as well as the question of the purity of enzyme preparations in isolated systems. It is clear that studies should be performed using nucleic acids, DNA and RNA polymerases and glycosaminoglycans in concentrations comparable to those claimed to exist in the nucleii before any biological structure-function significance can be placed on these effects.

There has been some circumstantial evidence in favour of a role of a 'heparin-like polysaccharide' in the transition of material-gene-directed development to embryonic genome-dependent differentiation of the sea urchin *in vivo*. When sea urchins are grown in sulphate-deficient sea water there development is not carried on beyond the blastula stage. Sulphate ions are taken up by the embryo in the beginning of gastrulation (Kinoshita, 1971, 1974) and become incorporated into a heparin-like polysaccharide. This polysaccharide is thought to be of cytoplasmic origin and subsequently transferred to the nucleus and to interact with template active parts of the chromatin. This interaction may then be responsible for changes in chromatin morphology and ultimately derepression of genes and enhanced RNA synthesis, as observed in model experiments on the addition of heparin or sea urchin polysaccharide to isolated nucleii obtained from pregastrular embryos but not for postgastrular embryos (Kinoshita, 1971, 1974, 1976).

6.6.2 Protein synthesis

It has been observed that heparin is a potent and specific inhibitor of protein synthesis, as measured by labelled amino acid incorporation into proteins in a cell free system (Waldman and Goldstein, 1973ab). On the addition of 4–5 μg/ml heparin to 20 μg/ml rabbit globin mRNA, a 50% inhibition of globin mRNA translation was observed. The effect appears quite specific for heparin in that other glycosaminoglycans i.e. chondroitin-4 and -6 sulphate, keratan sulphate, dermatan sulphate, heparan sulphate and hyaluronate, or the polysaccharide dextran sulphate were without significant effect (Waldman and Goldstein, 1973b; Waldman *et al.*, 1974; Gressner and Greiling, 1977). This pattern of inhibition would appear to rule out a general polyelectrolyte effect arising from line charge density considerations but would suggest a specific sequence of disaccharide units in heparin responsible for activity. Partially de-N-sulphated heparin (5 out of 17 N-sulphate groups removed) retained activity while having lost its anticoagulant properties, whereas partially de N- and O-sulphated heparin (corresponding to the loss of approximately 4.5 O-sulphate groups and 9 N-sulphate groups) had little activity (Waldman *et al.*, 1974; Goldstein *et al.*, 1975).

The mechanism of heparin action in this system is not clearly understood, as effects on both initiation (Waldman *et al.*, 1975) and elongation factors (Slobin, 1976) have been proposed. Heparin may interfere with the ribosomal attachment sites of the tRNA-amino acid-ribosome-peptide ternary complex (Slobin, 1976; Gressner and Greiling, 1977). Heparin-linked Sepharose columns have been used to isolate peptide initiation factors (Waldman *et al.*, 1975), elongation factors (Slobin, 1976) and all factors required for cell-free protein synthesis (Hradec and Dušek, 1978). It remains to be shown whether heparin is able to exert its protein synthesis inhibitory effects *in vivo* or even affect synthesis of heparin proteoglycan protein core.

REFERENCES

Akasaki, M., Kawasaki, T. and Yamashina, I., *FEBS Lett.* **59**, 100 (1975).

Amsterdam, A., Reches, A., Amir, Y., Mintz, Y. and Salomon, Y., *Biochim. Biophys. Acta* **544**, 273 (1978).

Anderson, P., Röhlich, P., Slorach, S. A. and Uvnäs, B., *Acta Physiol. Scand.* **91**, 145 (1974).

Anderson, P., Slorach, S. A. and Uvnäs, B., *Acta Physiol. Scand.* **88**, 359 (1973).

Ansevin, A. Y., MacDonald, K. K., Smith, C. E. A. and Hnilica, L. S., *J. Biol. Chem.* **250**, 281 (1975).

Arnold, E. A., Yawn, D. H., Brown, D. G., Wyllie, R. C. and Coffey, D. S., *J. Cell. Biol.* **53**, 737 (1972).

Augusti-Tocco, G. and Chiarugi, V. P., *Cell Differ.* **5**, 161 (1976).

Austen, K. F. and Humphrey, J. H., *Adv. Immunol.* **3**, 1 (1963).

Avila, J. L., *Biochem. J.* **171**, 489 (1978).

Avila, J. L. and Convit, J., *Biochem. J.* **152**, 57 (1975).

Avila, J. L. and Convit, J., *Biochem. J.* **160**, 129 (1976).

Berg, N. B., *J. Cell Sci.* **31**, 199 (1978).

Berg, N. B. and Young, R. W., *J. Cell Biol.* **50**, 469 (1971).

Bergqvist, U., Samuelsson, G. and Uvnäs, B., *Acta Physiol. Scand.* **83**, 362 (1971).

Berlin, G. and Enerbäck, L., *J. Histochem. & Cytochem.* **26**, 14 (1978).

Berlowitz, L., Kitchen, R. and Pallota, D., *Biochim. Biophys. Acta* **262**, 160 (1972).

Bhavanandan, V. P. and Davidson, E. A., *Proc. Natl. Acad. Sci. USA* **72**, 2032 (1975).

Blaschke, E., Bergqvist, U. and Uvnäs, B., *Acta Physiol. Scand.* **97**, 110 (1976).

Bornstein, P., Duksin, D., Balian, G., Davidson, J. M. and Crouch, E., *Ann. New York Acad. Sci.* **312**, 93 (1978).

Branford-White, C. J., *Experentia* **34**, 1036 (1978a).

Branford-White, C. J., *Arch. Oral Biol.* **23**, 1141 (1978b).

Brennessel, B. A., Buhrer, D. P. and Gottleib, A. A., *Anal. Biochem.* **87**, 411 (1978).

Brimacombe, J. S. and Weber, J. M., *Mucopolysaccharides. Chemical Structure, Distribution and Isolation*, Elsevier, Amsterdam (1964).

Buonassisi, V., *Exp. Cell Res.* **76**, 363 (1973).
Buonassisi, V. and Ozzello, L., *Cancer Res.* **33**, 874 (1973).
Buonassisi, V. and Root, M., *Biochim. Biophys. Acta* **385**, 1 (1975).
Burson, S. L., Fahrenbach, M. J., Frommhagen, L. J., Riccardi, B. A., Brown, R. A., Brockman, J. A., Lewry, H. V. and Stokstad, E. L. R., *J. Am. Chem. Soc.* **78**, 5874 (1956).
Cássaro, C. M. F. and Dietrich, C. P., *J. Biol. Chem.* **252**, 2254 (1977).
Chambon, P., Ramuz, M., Mandel, P. and Daly, J., *Biochim. Biophys. Acta* **157**, 504 (1968a).
Chambon, P., Karon, H., Ramuz, M. and Mandel, P., *Biochim. Biophys. Acta* **157**, 520 (1968b).
Chandrasekaran, E. V., Spolter, L. and Marx, W., *Prep. Biochem.* **5**, 281 (1975).
Chiarugi, V. P., *Exp. Cell Biol.* **44**, 251 (1976).
Chiarugi, V. P. and Dietrich, C. P., *J. Cell. Physiol.* **99**, 201 (1979).
Chiarugi, V. P. and Vannucchi, S., *J. Theor. Biol.* **61**, 459 (1976).
Choi, Y. C., Swack, N. S. and Hsuing, G. D., *Proc. Soc. Exp. Biol. Med.* **157**, 569 (1978).
Cifonelli, J. A. and King, J., *Biochim. Biophys. Acta* **215**, 273 (1970).
Cifonelli, J. A. and Mathews, M. B., *Connect. Tiss. Res.* **1**, 121 (1972).
Clementi, F. and Meldolesi, J., *J. Cell Biol.* **65**, 88 (1975).
Cohn, R. H., Cassiman, J. J. and Bernfield, M. R., *J. Cell Biol.* **71**, 280 (1976).
Comper, W. D. and Laurent, T. C., *Physiol. Rev.* **58**, 255 (1978).
Conrad, G. W. and Hart, G. W., *Dev. Biol.* **44**, 253 (1975).
Cook, R. T. and Aikawa, M., *Exp. Cell Res.* **78**, 257 (1973).
Coupar, B. E. H. and Chesterton, C. J., *Eur. J. Biochem.* **79**, 525 (1977).
Di Cioccio, R. A. and Srivastava, B. I. S., *Cancer Res.* **38**, 2401 (1978).
Dietrich, C. P. and De Oca, H. M., *Proc. Soc. Exp. Biol. Med.* **134**, 955 (1970).
Dietrich, C. P., Sampaio, L. O. and Toledo, O. M. S. *Biochem. Biophys. Res. Comm.* **71**, 1 (1976).
Dietrich, C. P., Sampaio, L. O., Toledo, O. M. S. and Cássaro, C. M., *Biochem. Biophys. Res. Commun.* **75**, 329 (1977).
Dorfman, A. and Ho, P.-L., *Proc. Natl. Acad. Sci. USA* **66**, 495 (1970).
Dorfman, A. and Matalon, R., *Proc. Natl. Acad. Sci. USA* **73**, 630 (1976).
Endo, M. and Yosizawa, Z., *J. Biochem.* **79**, 1 (1976).
Enerbäck, L., *Histochem.* **42**, 301 (1974).
Enerbäck, L. and Mellblom, L., *Cell & Tissue Res.* **187**, 367 (1978).
Everitt, M. T. and Neurath, H., *Biochimie* **61**, 653 (1979).
Fabian, I., Bleiberg, I. and Aronson, M., *Biochim. Biophys. Acta* **544**, 69 (1978).
Fabry, M., Sümegi, J. and Venetianer, P., *Biochim. Biophys. Acta* **435**, 228 (1976).
Ferencz, A. and Seifart, K. H., *Eur. J. Biochem.* **53**, 605 (1975).
Fillion, G., Slorach, S. A. and Uvnäs, B., *Acta Physiol. Scand.* **78**, 547 (1970).
Fromme, H. G., Buddecke, E., von Figura, K. and Kresse, H., *Exp. Cell. Res.* **102**, 445 (1976).
Frommhagen, L. H., Fahrenbach, M. J. and Brockman, J. A., Jr., *Proc. Soc. Exp. Biol. Med.* **82**, 280 (1953).
Funakoshi, I., Nakada, H. and Yamashina, I., *J. Biochem.* **76**, 319 (1974).

Furukawa, K. and Terayama, H., *Biochim. Biophys. Acta* **499**, 278 (1977).

Gebauer, H., Lindner, H. R. and Amsterdam, A., *Biol. Reprod.* **18**, 350 (1978).

Giannattasio, G. and Zanini, A., *Biochim. Biophys. Acta* **439**, 349 (1976).

Glick, D., von Redlick, D. and Diamant, B., *Biochem. Pharm.* **16**, 553 (1967).

Glimelius, B., Busch, C. and Höök, M., *Thromb. Res.* **12**, 773 (1978a).

Glimelius, B., Norling, B., Westermark, B. and Wasteson, Å., *Biochem. J.* **172**, 443 (1978b).

Goldstein, D. J. and Horobin, R. W., *Histochem. J.* **6**, 157, (1974).

Goldstein, J., Waldman, A. A. and Marx, G., in *Heparin: Structure, Function and Clinical Aspects*, R. A. Bradshaw and S. Wessler (eds.), Plenum Press, New York (Adv. Exp. Med. Biol. Vol 52) p289 (1975).

Gressner, A. M. and Greiling, H., *Hoppe-Seyler's Z. Physiol. Chem.* **358**, 69 (1977).

Gustafsson, B. and Enerbäck, L., *J. Histochem. & Cytochem.* **26**, 47 (1978).

Hart, G. W., *J. Biol. Chem.* **251**, 6513 (1976).

Hata, R. and Nagai, Y., *Biochim. Biophys. Acta* **304**, 408 (1973).

Hauser, P. and Vaes, G., *Biochem. J.* **172**, 275 (1978).

Helander, H. F. and Bloom, G. D., *J. Microsc.* **100**, 315 (1974).

Hiebert, L. M. and Jaques, L. B., *Thromb. Res.* **8**, 195 (1976a).

Hiebert, L. M. and Jaques, L. B., *Artery* **2**, 26 (1976b).

Higginbotham, R. D. and Dougherty, F. T., *Reticuloendothelial Soc. Bull.* **2**, 27 (1956).

Hildebrand, C. E., Gurley, L. R., Tobey, R. A. and Walters, R. A., *Biochim. Biophys. Acta* **477**, 295 (1977).

Hildebrand, C. E. and Tobey, R. A., *Biochem. Biophys. Res. Comm.* **63**, 134 (1975).

Horner, A. A., in *Heparin: Structure, Function and Clinical Implication*, R. A. Bradshaw and S. Wessler (eds.), Plenum Press, New York (Adv. Exp. Med. Biol. Vol. 52) p85 (1975).

Horner, A. A., *Fed. Proc.* **36**, 35 (1977).

Hradec, J. and Dušek, Z., *Biochem. J.* **172**, 1 (1978).

Hurst, R. E., Nakamura, N. and West, S. S., *Prep. Biochem.* **8**, 37 (1978).

Hynes, R. O., *Biochim. Biophys. Acta* **458**, 73 (1976).

Jamieson, J. D. and Palade, G. E., in *International Cell Biology 1976–1977*, B. R. Brinkley and K. R. Porter (eds.), Rockefeller University Press, p308 (1977).

Janakidevi, K., *Exp. Cell. Res.* **112**, 345 (1978).

Jaques, L. B., *Gen. Pharmac.* **6**, 235 (1975).

Jaques, L. B., Mahadoo, J. and Riley, J. F., *Lancet* **1**, 411 (1977).

Jaques, L. B. and Waters, E. T., *J. Physiol.* **99**, 454 (1941).

Jilek, F. and Hörmann, H., *Hoppe-Seyler's, Z. Physiol. Chem.* **360**, 597 (1979).

Johnston, L. S., Keller, K. L. and Keller, J. M., *Biochim. Biophys. Acta* **583**, 81 (1979).

Jorpes, J. E., *Heparin*, 2nd Ed., Oxford University Press, London (1946).

Kanwar, Y. S. and Farquhar, M. G., *Proc. Natl. Acad. Sci. USA* **76**, 1303 (1979).

Kawiak, J., Vensel, W. H., Komender, J. and Barnard, E. A., *Biochim. Biophys. Acta* **235**, 172 (1971).

Keller, K. L., Underhill, C. B. and Keller, J. M., *Biochim. Biophys. Acta* **540**, 431 (1978).

Kelley, R. O., Palmer, G. C., Crissman, H. A. and Nilson, J. H., *J. Cell. Sci.* **28**, 237 (1977).
Kent, P. W., Hickens, M. and Ward, P. F. V., *Biochem. J.* **68**, 568 (1958).
Kiernan, J. A., *J. Anat.* **121**, 303 (1976).
Kinoshita, S., *Exp. Cell. Res.* **64**, 403 (1971).
Kinoshita, S., *Exp. Cell. Res.* **85**, 31 (1974).
Kinoshita, S., *Exp. Cell Res.* **102**, 153 (1976).
Kitzis, A., Defer, N., Dastugue, B., Sabatier, M. M. and Kruh, J., *FEBS Lett.* **66**, 336 (1976).
Kjellén, L., Oldberg, Å., Rubin, K. and Höök, M., *Biochem. Biophys. Res. Commun.* **74**, 126 (1977).
Kobiler, D. and Barondes, S. M., *FEBS Lett.* **101**, 257 (1979).
Kobayashi, Y., *Arch. Biochem. Biophys.* **96**, 20 (1962).
Kraemer, P. M., *Biochemistry* **10**, 1437 (1971a).
Kraemer, P. M., *Biochemistry* **10**, 1445 (1971b).
Kraemer, P. M., *Biochem. Biophys. Res. Comm.* **78**, 1334 (1977).
Kraemer, P. M. and Tobey, R. A., *J. Cell Biol.* **55**, 713 (1972).
Kraemer, R. J. and Coffey, D. S., *Biochim. Biophys. Acta* **224**, 568 (1970a).
Kraemer, R. J. and Coffey, D. S., *Fed. Proc.* **29**, 896 (1970b).
Kresse, H., von Figura, K., Buddecke, E. and Fromme, H. G., *Hoppe-Seyler's Z. Physiol. Chem.* **356**, 929 (1975a).
Kresse, H., Tekolf, W., von Figura, K. and Buddecke, E., *Hoppe-Seyler's Z. Physiol. Chem.* **356**, 943 (1975b).
Kronquist, K. E., Elmahdy, A. and Ronzio, R. A., *Arch. Biochem. Biophys.* **182**, 188 (1977).
Kupchella, C. E. and Steggerda, F. R., *Trans. N.Y. Acad. Sci.* **34**, 351 (1972).
Lagunoff, D., *Biochem. Pharmacol.* **21**, 1889 (1972).
Lagunoff, D., *Biochemistry* **13**, 3982 (1974).
Lagunoff, D., Phillips, M. T., Iresi, O. A. and Benditt, E. P., *Lab. Invest.* **13**, 1331 (1964).
Lindahl, U., Pertoft, H. and Seljelid, R., *Biochem. J.* **182**, 189 (1979).
Linker, A. and Hovingh, P., *Carbohydr. Res.* **29**, 41 (1973).
Lis, D. and Monis, B., *Experientia* **34**, 693 (1978).
Mäenpää, P. H., *Biochim. Biophys. Acta* **498,** 294 (1977).
Margolis, R. K., Crockett, C. P., Kiang, W.-L. and Margolis, R. U., *Biochim. Biophys. Acta* **451**, 465 (1976).
Margolis, R. U. and Margolis, R. K., *Int. J. Biochem.* **8**, 85 (1977).
Margolis, R. U. and Margolis, R. K., *Biochem. Pharmacol.* **22**, 2195 (1973).
McKusick, V. A., Neufeld, E. F. and Kelly, T. E., in *The Metabolic Basis of Inherited Disease*, J. B. Stanbury, J. B. Wyngaarden and D. S. Frederickson (eds.), McGraw-Hill, New York, p1282 (1978).
Meier, S. and Hay, E. D., *Proc. Natl. Acad. Sci. USA* **71**, 2310 (1974).
Miller, G. J., Berlowitz, L. and Regelson, W., *Exp. Cell. Res.* **71**, 409 (1972).
Moran, N. C., Uvnäs, B. and Westerholm, B., *Acta Physiol. Scand.* **56**, 26 (1962).
Morgan, R. E., Moore, J. S. and Phillips, G. O., *J. Histochem. Cytochem.* **20**, 831 (1972).

Murata, K., *Clin. Chim. Acta* **63**, 157 (1975).
Murata, K., Ogura, T. and Okuyama, T., *Connect. Tiss. Res.* **2**, 101 (1974).
Nakada, H., Funakoshi, I. and Yamashina, I., *J. Biochem.* **78**, 863 (1975).
Neufeld, E. F., Lim, T. W. and Shapiro, L., *J. Ann. Rev. Biochem.* **44**, 357 (1975).
Öbrink, B., *Eur. J. Biochem.* **233**, 387 (1973).
Oldberg, Å., Höök, M., Öbrink, B., Pertoft, H. and Rubin, K., *Biochem. J.* **164**, 75 (1977).
Oldberg, Å., Kjellén, L. and Höök, M., *J. Biol. Chem.* **254**, 8505 (1979).
Olivecrona, T., Bengtsson, G., Markland, S.-E., Lindahl, U. and Höök, M., *Fed. Proc.* **36**, 60 (1977).
Orenstein, N. S., Galli, S. J., Dvorak, A. M., Silbert, J. E. and Dvorak, H. F., *J. Immol.* **121**, 586 (1978).
Padawer, J., *Am. J. Anat.* **141**, 299 (1974).
Pomerai, D. I. de, Chesterton, C. J. and Butterworth, P. H., *FEBS Lett.* **42**, 149 (1974).
Pritchard, D. G., Halpern, R. M. and Smith, R. A., *Biochim. Biophys. Acta* **228**, 127 (1971).
Radhakrishnamurthy, B., Ruiz, H. A. and Berenson, G. S., in *Atherosclerosis. Metabolic, Morphologic and Clinical Aspects*, G. W. Manning and M. D. Haust (eds.), Plenum Press, New York and London, p160 (1978a).
Radhakrishnamurthy, B., Ruiz, H. A., Dalferes, E. R. Jr., Friedman, M., Seethanathan, P. and Berenson, G. S., *Athero-sclerosis* **29**, 25 (1978b).
Rahemtulla, F. and Løvtrup, S., *Comp. Biochem. Physiol.* **508**, 631 (1975).
Reggio, H. A. and Dagorn, J. C., *J. Cell. Biol.* **78**, 951 (1978).
Reggio, H. A. and Palade, G. E., *J. Cell Biol.* **77**, 288 (1978).
Riley, J. F., *Can. J. Biochem. Physiol.* **39**, 633 (1961).
Roblin, R., Albert, S. O., Gelb, H. A. and Black, P. H., *Biochemistry* **14**, 347 (1975).
Rollins, B. J. and Culp, L. A., *Biochemistry* **18**, 141 (1979).
Saiga, H. and Kinoshita, S., *Exp. Cell Res.* **102**, 143 (1976).
Sakamoto, S., Sakamoto, M., Goldhaber, P. and Glimcher, M. J., *Biochim. Biophys. Acta* **385**, 41 (1975).
Salomon, Y., Amir, Y., Azulai, R. and Amsterdam, A., *Biochim. Biophys. Acta* **544**, 262 (1978).
Salomon, Y. and Amsterdam, A., *FEBS Lett.* **83**, 263 (1977).
Satoh, C., Banks, J., Horst, P., Kreider, J. W. and Davidson, E. A., *Biochemistry* **13**, 1233 (1974).
Schäfer, R., Zillig, W. and Zechel, K., *Eur. J. Biochem.* **33**, 207 (1973).
Schaffrath, D., Stuhlsatz, H. W. and Greiling, H., *Hoppe-Seyler's Z. Physiol. Chem.* **357**, 499 (1976).
Schiller, S. and Dorfman, A., *Biochim. Biophys. Acta* **31**, 278 (1959).
Scott, J. E., *Biochem. Soc. Trans.* **1**, 787 (1973).
Scott, J. E. and Willet, I. H., *Nature* **209**, 985 (1966).
Seethanathan, P., Radhakrishnamurthy, B., Dalferes, E. R. Jr., and Berenson, G. S., *Respir. Physiol.* **24**, 347 (1975).
Seifart, K. H., Ferencz, A. and Benecke, B. J., in *Control of Transcription and Translation in Eucaryotes*, E. K. F. Bautz (ed.), 24th Mosbacher Colloquium, Springer-Verlag, Heidelberg, New York, p595 (1973).

Seki, S. and Oda, T., *Biochim. Biophys. Acta* **479**, 391 (1977).

Serafini-Fracassini, A., Brandford-White, C. J. and Hunter, J. C., *FEBS Lett.* **32**, 116 (1973).

Shanberge, J. N., Kambayashi, J. and Nakagawa, M., *Thromb. Res.* **9**, 595 (1976).

Singer, S. J. and Nicolson, G. L., *Science* **175**, 720 (1972).

Sjöberg, I. and Fransson, L.-Å., *Biochem. J.* **167**, 383 (1977).

Slobin, L. I., *Biochem. Biophys. Res. Commun.* **73**, 539 (1976).

Slorach, S. A., *Acta Physiol. Scand.* **82**, 91 (1971).

Smith, M. R. and Cook, R. T., *Exp. Cell Res.* **110**, 15 (1977).

Stathakis, N. E. and Mosesson, M. W., *J. Clin. Invest.* **60**, 855 (1977).

Stein, G. S., Roberts, R. M., Davis, J. L., Head, W. J., Stein, J. L., Thrall, C. L., Veen, J. V. and Welch, D. W., *Nature* **258**, 639 (1975).

Stevens, R. L., Colombo, M., Gonzales, J. J., Hollander, W. and Schmid, K., *J. Clin. Invest.* **58**, 470 (1976).

Stone, A. L., in *Structure and Stability of Biological Macromolecules* (Biological Macromolecules Series, Vol. 2), G. Fasman, and S. Timasheff (eds.), Marcel Dekker, New York, p353 (1969).

Sue, T. K. and Jaques, L. B., *Proc. Soc. Exp. Biol. Med.* **146**, 1006 (1974).

Sue, T. K. and Jaques, L. B., *Atherosclerosis* **25**, 137 (1976).

Takata, K. and Terayama, H., *Biochim. Biophys. Acta* **500**, 333 (1977).

Tammi, M., Seppälä, P. O., Lehtonen, A. and Möttönen, M., *Atherosclerosis* **29**, 191 (1978).

Tas, J., *Histochem. J.* **7**, 1 (1975).

Tas, J., *Histochem. J.* **9**, 205 (1977).

Tas, J. and Berndsen, R. G., *J. Histochem. Cytochem.* **25**, 1058 (1977).

Taylor, S. R. and Cook, R. T., *FEBS Lett.* **78**, 321 (1977).

Thomas, L. J. Jr., *Biol. Bull.* **101**,, 230 (1951).

Thomas, L. J. Jr., *Biol. Bull.* **106**, 129 (1954).

Toledo, O. M. S. and Dietrich, C. P., *Biochim. Biophys. Acta* **498**, 114 (1977).

Underhill, C. B. and Keller, J. M., *J. Cell Physiol.* **89**, 53 (1976).

Underhill, C. B. and Keller, J. M., *J. Cell Physiol.* **90**, 53 (1977).

Uvnäs, B., in *Handbook of Exp. Pharmacol.*, Springer-Verlag, Berlin–Heidelberg, Vol. XVIII/2 (1977).

Uvnäs, B. and Åborg, C.-H., *Acta Physiol. Scand.* **96**, 512 (1976).

Uvnäs, B., Åborg, C.-H. and Bergendorff, A., *Acta Physiol. Scand. Suppl.* No. **336**, 1 (1970).

Uvnäs, B., Åborg, C.-H. and Bergqvist, U., *Acta Physiol. Scand.* **93**, 401 (1975).

Vaes, G., Eeckhout, Y., Lenaers-Claeys, G., Francois-Gillet, C. and Dreutz, J.-E., *Biochem. J.* **172**, 261 (1978).

Vaheri, A., Ruoslahti, E., Westermark, B. and Ponten, J., *J. Exp. Med.* **143**, 64 (1976).

Waldman, A. A. and Goldstein, J., *Biochim. Biophys. Acta* **331**, 243 (1973a).

Waldman, A. A. and Goldstein, J., *Biochemistry* **12**, 2706 (1973b).

Waldman, A. A., Marx, G. and Goldstein, J., *Biochim. Biophys. Acta* **343**, 324 (1974).

Waldman, A. A., Marx, G. and Goldstein, J., *Proc. Natl. Acad. Sci. USA* **72**, 2352 (1975).

Walters, R. A. and Hildebrand, C. E., *Biochim. Biophys. Acta* **407**, 120 (1975).
Walter, G., Zillig, W., Palm, P. and Fuchs, E., *Eur. J. Biochem.* **3**, 194 (1967).
Warnick, C. T. and Lazarus, H. M., *Nucleic Acids Res.* **2**, 735 (1975).
Wasteson, Å., Glimelius, B., Busch, C., Westermark, B., Heldin, C.-H. and Norling, B., *Thromb. Res.* **11**, 309 (1977).
Werb, Z. and Burleigh, M. C., *Biochem. J.* **137**, 373 (1974).
Wilander, O., *Skand. Arch. Physiol.* S15 (1938).
Wilhelm, D. L., Young, L. C. J. and Watkins, S. G., *Agents and Actions* **8**, 146 (1978).
Wusteman, F. S., *Experientia* **28**, 887 (1972).
Yamada, K. M., Yamada, S. S. and Pastan, I., *Proc. Natl. Acad. Sci. USA* **72**, 3158 (1975).
Yamamoto, K. and Terayama, H., *Cancer Res.* **33**, 2557 (1973).
Yosizawa, Z., *Biochem. Biophys. Res. Comm.* **16**, 336 (1964).
Young, I. J. and Custod, J. T., *J. Neurochem.* **19**, 923 (1972).
Young, R. W., *J. Cell. Biol.* **57**, 175 (1973).
Yurt, R. and Austen, K. F., *J. Exp. Med.* **146**, 1405 (1977).
Yurt, R. W., Leid, R. W. Jr., Austen, K. F. and Silbert, J. E., *J. Biol. Chem.* **252**, 518 (1977a).
Yurt, R. W., Leid, R. W. Jr., Spragg, J. and Austen, K. F., *J. Immunol.* **118**, 1201 (1977b).
Zillig, W., Zechel, K., Rabussay, D., Schacher, M., Sethi, V. S., Palm, P., Heil, A. and Seifart, W., *Cold Spring Harbor Symp. Quant. Biol.* **35**, 47 (1971).

CHAPTER 7

Extracellular interactions

7.1 BLOOD HEPARIN?

The presence of heparin in blood has been and remains a controversial issue. In view of its highly active pharmacological properties, such as its anticoagulant, antilipemic and anticomplement activity, it has been assumed that exogenous heparin remaining in the vascular compartment is responsible for this activity. However, there is little evidence to ascertain the locality of heparin from where it manifests its pharmacological activity. Is its site of activity in blood or is exogenous heparin in blood in equilibrium with particular localization sites from where it exerts its activity? In translating the properties of exogenous heparin to those of endogenous heparin, studies have focused on establishing whether endogenous heparin exists in blood. These studies have met with ambiguous results. It appears that in most animals there is little, if any, endogenous heparin in the blood. It remains questionable whether endogenous heparin has the pharmacological activity of exogenous material *in vivo*.

It has been known for many years that anaphylactic shock in dogs causes degranulation of mast cells in the liver with subsequent release of heparin into the blood (Jaques and Waters, 1941). However, Riley (1963) has pointed out that the dog is exceptional in releasing heparin from its mast cells so readily and that in other mammals this reaction may not occur to any significant extent (see also Section 6.2.1). A number of other early studies by Engelberg and co-workers (Engelberg, 1977 and references therein) have provided circumstantial evidence for the existence of heparin in blood; mainly sought through criteria involving anticoagulant activity. Their plasma extracts, possessing heparin-like activity, have yet to be chemically analyzed for heparin content.

The importance of endogenous heparin localization other than heparin being in the free, soluble form in blood has been demonstrated by Horner (1974). Endogenous [^{35}S]-labelled heparin was prepared by intraperitoneal injection of [^{35}S]-sulphate into rats and the blood was collected 18 hours later. Commercial heparin carrier was added to the isolated fractions of red blood cells, platelets, plasma and buffy layer, subsequent to treatment with pronase. The heparins were precipitated with cetylpyridinium chloride and fractionated by ion-exchange chromatography. It was found that 50% (in terms of [^{35}S]-sulphate radioactivity) of the heparin was in the platelet product, 30% was recovered from the red cells, 10% from the buffy layer and 10% from the plasma. Therefore, 90% of the blood heparin was associated with formed elements. It was estimated that the blood heparin concentration was in the range of 5.5–42 μg/ml. It is also of interest that platelet heparin was of low molecular weight.

More recently, Jacobsson and Lindahl (1979) have developed a micro-scale determination of heparin in human blood plasma involving extraction with water-saturated phenol and recovery from the aqueous phase by ion-exchange chromatography on DEAE-cellulose. The heparin content of the product was determined by a colourimetric antithrombin-activation assay. The amount of endogenous heparin, if any, in human blood was calculated to be less than 2 ng per ml.

It is known that heparin binds to many proteins in plasma although the nature of these interactions and their effect on heparin action is not clearly understood. The weak binding of heparin to plasma components α-, γ-, and β-globulin (Takeda and Kobayashi, 1977), cold-insoluble globulin (Stathakis and Mosesson, 1977) and fibrinogen (Marciniak, 1974; Takeda and Kobayashi, 1977; Kudryashov *et al.*, 1977) but not albumin (Marciniak, 1974; Blaskó *et al.*, 1977) has been demonstrated. The binding of heparin to lipoproteins (Section 7.5) and haemoglobin (Amiconi *et al.*, 1977) has also been demonstrated. Gel chromatographic studies of plasma proteins admixed with heparin (as determined by dye binding (Marciniak, 1974) or intrinsic [^{35}S]-

labelling of heparin (Takeda and Kobayashi, 1977)) have shown that heparin binding to antithrombin III is small (see, however, Shanberge *et al.*, 1976), although the heparin retains its anticoagulant activity when bound to plasma proteins. A general conclusion arising from these studies is that heparin binds to proteins other than antithrombin III in plasma, because of the much greater quantity of the former. These studies do suggest that only a small fraction of heparin *in vivo* may be active and/or the binding to plasma proteins may serve as a vehicle for its deposition at a site where it manifests anticoagulant activity.

Although endogenous heparin has only been identified essentially in mast cell granules, its existence *in vivo* on various cell surfaces seems possible in view of the results of binding of exogenous heparin to cell surfaces of platelets (Shanberge *et al.*, 1976), endothelial cells (Hiebert and Jaques, 1976ab; Glimelius *et al.*, 1978) and rat liver cells (Kjellén *et al.*, 1977). It may impart on these regions, where blood–surface interfaces exist, protection against thrombus formation and proteolytic attack. The presence of heparan sulphate on the surface of platelets and endothelial cells may also contribute to the antithrombitic properties of these interfaces. This may explain the difficulty in obtaining evidence for the existence of heparin in blood in that, on its release from mast cells, it may be taken up locally by various surfaces for control of local antithrombitic properties.

7.2 ANTICOAGULANT ACTIVITY OF HEPARIN

7.2.1 The coagulation system

The coagulation system is part of the continuum of host response to injury and is thus intimately involved with the kinin, complement and fibrinolytic systems. Heparin is known to affect all these systems and its action will be considered in the following sections.

Blood coagulation occurs as a series of complex steps, which terminate in the formation of a fibrin clot. It can be brought about when blood comes in contact with non-endothelial surfaces (often referred to as a 'foreign' surface) such as collagen *in vivo* or glass *in*

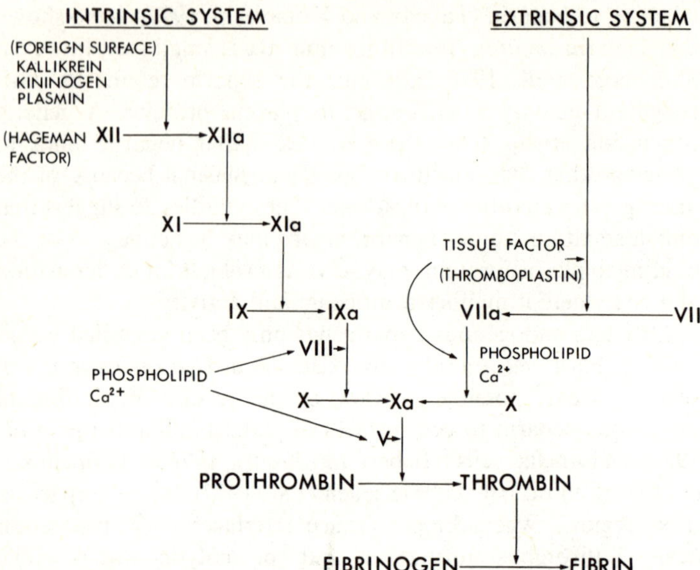

FIGURE 7.1 Schematic representation of the coagulation system (see text for details).

vitro. A simplified representation of the blood clotting cascade is given in Figure 7.1. It consists of an intrinsic pathway, which is mediated entirely by components found in plasma and is a relatively slow process compared to the extrinsic pathway which is initiated by lipoproteins (tissue thromboplastin) found in various tissues. The coagulation cascade is composed of a series of linked proteolytic reactions. At each stage of this mechanism a parent zymogen is converted to a corresponding serine protease, which catalyzes a subsequent zymogen–protease transition. Conversion of the zymogen to an active enzyme by peptide bond hydrolysis is usually exergonic, so contributing to the irreversible nature of the coagulation cascade. Both pathways converge to produce the active proteolytic enzyme Factor Xa, which in turn catalyzes the conversion of prothrombin into thrombin by limited proteolysis. Thus, of the eleven proteins directly or indirectly involved in the conversion of prothrombin to thrombin, five proteins are ultimately activated to serine proteases (Factors XIIa, XIa, IXa, VIIa and Xa) and three proteins are cofactors for these proteolytic events (high molecular weight kininogen and Factors VIII and V). Thrombin is the end product of the coagulation cascade. This enzyme may then cleave two sets of arginine–glycine bonds in fibrinogen. This action releases fibrinopeptides A and B from the macromolecule and thus converts it to fibrin. Fibrin can then undergo a complex series of polymerization reactions that result in the formation of the thrombus.

Thrombin occupies a vital position in the haemostatic mechanism. It is intimately involved in other areas apart from directly catalyzing fibrin production. It potentiates the activity of factors VIIa, VIII, X and XIII in the clotting scheme. Thrombin is also able to produce platelet aggregation presumably after binding to a receptor on the platelet cell surface. It is clear then that inhibition of thrombin or prevention of its formation will effectively inhibit thrombus formation. Such neutralization is provided by a naturally occurring α_2-globulin, called antithrombin III, whose inhibitory effects are potentiated by heparin.

In terms of heparin anticoagulant activity it was known early that heparin requires a plasma cofactor (Howell, 1925; Brinkhous *et al.*, 1939). It was some 30 years later that Abildgaard (1968) isolated small amounts of a purified human plasma protein which, in the presence of heparin, rapidly neutralized thrombin activity. It is now recognized that there are at least three proteins which contribute to the progressive, irreversible destruction of thrombin, namely α_1-antitrypsin, α_2-macroglobulin (these are general protease inhibitors) and an α_2-globulin with more specific antithrombin activity, specifically potentiated by heparin, which is known as antithrombin III (which will be described simply as antithrombin). Of the total antithrombin activity of plasma, 80–90% is attributed to antithrombin (Rosenberg and Damus, 1973). The plasma antithrombin content is about 2 μM (Collen *et al.*, 1977). It circulates in the blood mainly in the free form with a half life of 2–8 days (Marciniak, 1974; Collen *et al.*, 1977). The molecular weights from human and bovine sources lie between 56,000–58,000 (Nordenman *et al.*, 1977).

Experiments *in vitro* have shown that antithrombin, by itself, may exert a slow progressive neutralization of thrombin

(Brinkhous *et al.*, 1939; Waugh and Fitzgerald, 1956; Monkhouse *et al.*, 1955; Abildgaard, 1968) and activated factor X (Xa) (Yin and Wessler, 1970; Seegers and Marciniak, 1962; Biggs *et al.*, 1970; Yin *et al.*, 1971). These studies have shown that in the presence of heparin the inhibition of these proteases is nearly instantaneous. Similar results have been shown for factor XIIa (Stead *et al.*, 1976), factor XIa (Damus *et al.*, 1973) and factor IXa (Rosenberg *et al.*, 1975). Furthermore, antithrombin slowly neutralizes the activity of each enzyme by the formation of a 1:1 enzyme–inhibitor complex. The addition of heparin results in a dramatic acceleration of complex formation without alteration in the stoichiometry of the reaction (Damus *et al.*, 1973; Highsmith and Rosenberg, 1974; Rosenberg *et al.*, 1975; Stead *et al.*, 1976). The only exception to the generalized view of heparin action on serine proteases of the coagulation cascade may be the inhibition of factor VIIa by antithrombin where the heparin effect appears minimal (Rosenberg, 1977). Other steps of the coagulation cascade mechanism that do not involve the action of serine proteases are only minimally affected by heparin.

It is now generally accepted that heparin functions as an anticoagulant by directly enhancing the inhibitory reaction which takes place between antithrombin and thrombin or factor Xa (Abildgaard, 1968; Rosenberg and Damus, 1973; Yin *et al.*, 1971; Gitel *et al.*, 1977). However, the contribution of heparin effects on the other serine proteases in the coagulation cascade is uncertain at the present time. In any case, the following sections will consider specifically the action of heparin on thrombin and factor Xa.

Before proceeding to these sections, it only needs to be mentioned that problems are associated in the field of assay methods for determining heparin and its anticoagulant activity. Two general types of assay measure the action of heparin on coagulation. First there are those determining the clotting of whole blood plasma. The inhibition in this type of system is not clearly understood. The second type of assay, developed more recently, involves the use of the rate of inactivation of a specific clotting protease.

The various tests yield different results, especially in terms of various heparin preparations. There has yet been no precise demonstration of the relation of the potency of heparin *in vitro* to exogenous activity *in vivo*. The various aspects of these tests will not be considered here, but have been discussed widely in terms of clinical management.

7.2.1.1 THROMBIN–ANTITHROMBIN INTERACTION

A number of difficulties in the study of this complex reaction have obviated a clear interpretation of its mechanism. These include the use of crude preparations of proteins, the variable activity of unfractionated heparin preparations and intrinsic differences associated with assays monitoring residual thrombin activity. It is only in the last few years that major advances in the understanding of the heparin action on potentiating antithrombin activity in neutralizing thrombin have been obtained. These studies have placed the heparin–antithrombin–thrombin interaction on a highly sophisticated level in terms of not only understanding heparin anticoagulant activity but polysaccharide–protein interactions in general. Even if the biological significance of the reaction in terms of endogenous heparin function is tenuous, it remains highly profitable to consider this reaction in some detail.

To demonstrate typical kinetics of this reaction Rosenberg and Damus (1973) have studied the inactivation of purified human thrombin with antithrombin in the presence and absence of heparin (Figure 7.2). If thrombin is added to buffer with or without heparin, no loss in enzymatic activity is noted. If thrombin is incubated for varing periods of time with antithrombin, a slow, progressive decline in the activity of the enzyme (as measured by coagulation or esterolytic assay) becomes apparent. If the thrombin is incubated with a mixture of heparin and antithrombin, its enzymatic activity is inhibited almost instantaneously. The equivalence point of this interaction occurs at near 1:1 molar ratio of enzyme to inhibitor whether in the presence or absence of heparin as determined by final concentrations of enzyme and inhibitor and by observation of the thrombin–antithrombin complex by analytical sodium dodecyl sulphate gel electrophoresis. Both antithrombin and thrombin migrate as single bands in this electrophoretic system with apparent molecular weights of 62,300 and 33,800 respectively. When these proteins are incubated

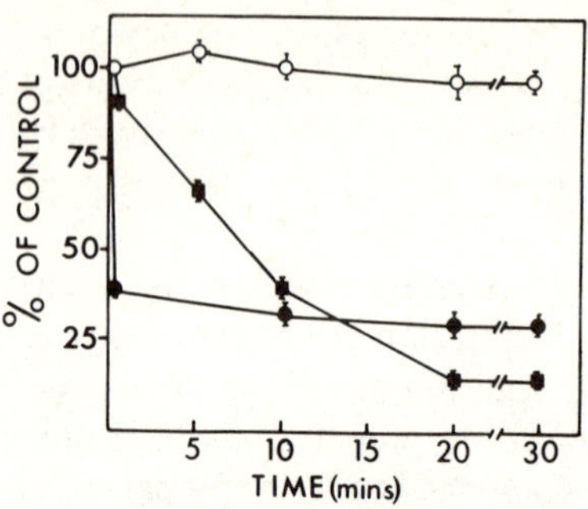

FIGURE 7.2 Inhibition of thrombin by antithrombin III in the presence and absence of heparin. Final concentrations of enzyme and inhibitor, 0.162 and 0.164 absorbance units/ml, respectively. Final concentration of added heparin, 12 u/ml (u = U.S.P. units of anticoagulant activity). Thrombin was measured by use of esterolytic assay using synthetic substrate tosyl-L-lysine methyl ester. ○, Thrombin + buffer; ■, antithrombin + thrombin; ●, antithrombin + heparin + thrombin; I, standard error of mean, as indicated on data points. (Adapted from Rosenberg and Damus, 1973 and Rosenberg, 1977; reproduced with permission of the publishers.)

together, both bands disappear over 5 minutes as enzyme activity is neutralized and results in the formation of a new molecular species of molecular weight 88,700.

The interaction of thrombin with antithrombin requires the presence of the active center serine of the enzyme and arginine residue(s) on the inhibitor, since chemical modification of either of these critical residues inhibits complex formation both in the presence and absence of heparin (Rosenberg and Damus, 1973) (see Table 7.2). Rosenberg and Damus (1973) suggested that a specific interaction occurs between the active center serine of thrombin and a unique arginine-x reactive site on antithrombin. This interaction was viewed to be analogous to trypsin–trypsin inhibitor systems where it was postulated (see for example Laskowski and Sealock, 1971) that the protease–protease inhibitor interaction results in the cleavage of a unique peptide bond (reactive site of the inhibitor) with formation of an acyl complex between the newly formed COOH-terminal residue of the inhibitor and the active site serine of the enzyme.

The thrombin–antithrombin complex so formed is a stable complex. It is not dissociated after treatment at 100°C (Rosenberg

TABLE 7.1
Binding Constants K_{ass} of Heparin to Antithrombin III

Heparin type	Method	K_{ass} (M^{-1})	Molar Stoichiometry (Hep/AT)	Solvent	Reference
H.A. (M = 6000)	F	$0.5\text{–}1.25 \times 10^7$	1.0	0.2MNaCl 0.01M Tris pH 7.5, 37°C	Jordan *et al.* (1979)
L.A. (M = 6000)	F	1×10^4	1.0	0.2M NaCl 0.01M Tris pH 7.5, 37°C	Jordan *et al.* (1979)
L.A. (M = 15,300)	U.V	$6\text{–}9 \times 10^4$	0.7–0.9	0.1M NaCl 0.05M Tris pH 7.4, 22°C	Nordenman and Björk (1978a)
H.A. (M = 15,300)	U.V.	$\geqslant 10^7$	0.85–0.9	0.1M NaCl 0.05M Tris pH 7.4, 22°C	Nordenman and Björk (1978a)
H.A. (M = 15,300)	C.D.	$\geqslant 10^7$	0.85–0.9	0.1M NaCl 0.05M Tris pH 7.4, 22°C	Nordenman and Björk (1978a)
H.A. (M = 15,300)	F	$8 \pm 3 \times 10^7$	0.8–1.0	0.1M NaCl 0.05M Tris pH 7.4, 22°C	Nordenman *et al.* (1978)
H.A. (M = 11,200)	F	2.3×10^6	—	0.2M glycine 0.05M Tris 0.03M NaCl pH 7.4, 25°C	Einarsson and Andersson (1977)
Heparin fractionated by gel chromatography (M = 10,700)	F	0.6×10^6 0.2×10^6 (2 binding sites)	—	0.2M glycine 0.05M Tris 0.03M NaCl pH 7.4, 25°C	Einarsson and Andersson (1977)

Abbreviations:
H.A. = High-affinity heparin for antithrombin III
L.A. = Low-affinity heparin for antithrombin III
F = Fluorescent spectroscopy
U.V. = Ultraviolet absorption difference spectroscopy
C.D. = Circular dichroism

TABLE 7.2
The Effect of Chemical Modification of Thrombin (T) and Antithrombin III (AT) on Their Interaction with Heparin

Protein	Chemical modifier	Modified amino acid	T activity	AT inhibition of T	Hep potentiation of AT inhibition of T	AT-Hep interaction	T-Hep interaction	T-Hep-AT interaction	Reference
Thrombin	diisopropylfluorophosphate	serine	I	—	—	—	—	NE	Rosenberg and Damus (1973); Owen (1975); Pomerantz and Owen (1978)
Thrombin	phenyl-methyl-sulphonyl fluoride	serine	I	—	—	—	NE	—	Nordenman and Björk (1978b)
Thrombin	proflavine dye	serine	I	—	—	—	NE	—	Li *et al.* (1974)
Thrombin	2,3-butanedione or 1,2-cyclohexanedione	5–6 arginine residues	variable	NE	I	—	—	—	Machovich (1975); Machovich *et al.* (1978); Pomerantz and Owen (1978)
Antithrombin III	2,3-butanedione or 1,2-cyclohexanedione	arginine	—	I	I	—	—	—	Rosenberg and Damus (1973)
Antithrombin III	0-methylisourea or 1,2-cyclohexabedione	lysine	—	NE	I	I	—	—	Rosenberg and Damus (1973)
Antithrombin III	N-bromosuccinamide	1–4 tryptophan residues	—	NE	I	—	—	—	Björk *et al.* (1979)

Abbreviations:
I = Inhibition
NE = No effect

and Damus, 1973) or by treatment with dodecyl sulphate (Rosenberg and Damus, 1973) or during gel chromatography in 6M HCl (Owen, 1975). The complex can be dissociated by treatment of the dodecyl sulphate–denatured complex with hydroxylamine and in dilute NaOH solutions at pH 12 (Owen, 1975). This evidence, together with the fact of the lack of a reaction between diisopropylphosphate–thrombin and the inhibitor (Rosenberg and Damus, 1973) suggests that a covalent ester linkage forms between the active site serine of the thrombin and a carboxyl group of the inhibitor (Rosenberg and Damus, 1973; Owen, 1975). It is viewed that covalent linkage of enzyme–inhibitor complex is irreversible and therefore will not only inhibit but essentially result in complete removal of enzyme activity from plasma.

While evidence for proteolytic modification of antithrombin by thrombin was indirectly demonstrated by Rosenberg and Damus (1973), it has only been recently established through the studies of Fish and Björk that this reaction does actually occur. They have demonstrated that in addition to the formation of an inactive, stable complex between thrombin and antithrombin, a large proportion of the antithrombin undergoes proteolysis by thrombin (Fish *et al.*, 1979). This modified antithrombin (AT_m) can only be detected by SDS gel electrophoresis under reducing and denaturing conditions of solutions of reacted thrombin and antithrombin. By this technique, the formation of AT_m was correlated qualitatively with the appearance of the antithrombin–thrombin complex and disappearance of free thrombin and antithrombin. Fish and Björk (1979) have also demonstrated its release from purified, stable, inactive complex by dissociation with hydroxylamine or ammonia in the presence or absence of sodium dodecyl sulphate or by prolonged treatment at 100°C in the presence of dodecyl sulphate. No evidence for release of unmodified antithrombin was observed. In urea/polyacrylamide gel electrophoresis of the modified antithrombin, only two polypeptide fragments were apparent with molecular weights of ~50,000 and 5,000 (Fish *et al.*, 1979).

Subsequently, Jörnvall *et al.* (1979) demonstrated that the modification by thrombin which produces these two species is at a cleavage site, an Arg–Ser bond, near the C-terminus of antithrombin. The cleavage liberates Ser-386 as the N-terminus of the small fragment and residue 385, which is Arg in human antithrombin (Petersen *et al.*, 1979). The two peptide chain fragments which result from this cleavage are held together by one or more disulphide bonds (Fish *et al.*, 1979).

A tentative scheme for the reaction of thrombin and antithrombin is as follows (Fish and Björk, 1979).

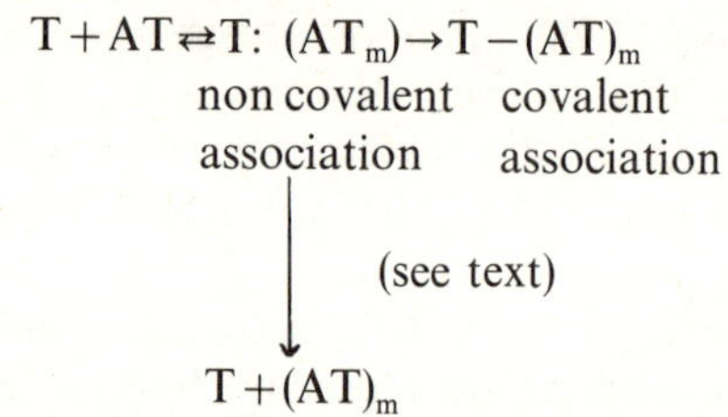

The reaction scheme accounts for an apparently unique feature of the antithrombin–thrombin reaction, i.e. simultaneous production of free modified antithrombin and inactive complex. Fish and Björk (1979) suggest that the initial stage of the reaction is characterized by antithrombin acting as a specific substrate for thrombin. Once the enzyme–inhibitor complex is formed, a specific bond of the inhibitor is cleaved. This is in contrast to a group of serine protease inhibitors from plants or animal tissues, such as the trypsin–trypsin inhibitor complex where evidence suggests that bond cleavage of the inhibitor does not occur but bonds of the tetrahedral intermediate type with the active site serine are formed. After proteolysis of antithrombin in the initial antithrombin–thrombin complex, two alternative routes become available for the modified antithrombin-molecule. In one of these, it may dissociate as a product from the enzyme-substrate complex in a reaction which produces the free modified inhibitor. The alternative route available to the proteolytically modified antithrombin is conformational rearrangement, which may involve the formation of a covalent acyl ester bond between enzyme and modified inhibitor, to yield an inactive antithrombin–thrombin complex. These two competing routes for modified antithrombin occur fairly rapidly but at comparable rates. This feature accounts for the observation that comparable amounts of both free modified antithrombin and inactive complex are formed concurrently during the antithrombin–thrombin reaction (Fish *et al.*, 1979).

The possible formation of free modified antithrombin, without concomitant neutralization of thrombin, is a most important feature of this reaction; particularly in relation to the heparin effect on this reaction to be discussed later. However, quantitative estimates for its formation on the reaction of thrombin with antithrombin have yet to be forthcoming. Fish *et al.* (1979) have obtained free modified antithrombin by thrombin treatment of antithrombin adsorbed to heparin-linked Sepharose; the modified antithrombin was released into solution by washing with 0.05M Tris HCl, 0.2M NaCl, pH 7.5 and purified by gel chromatography. This modified antithrombin showed identical electrophoretic behaviour on SDS gel electrophoresis to the modified antithrombin component in mixtures of thrombin and antithrombin. It had little, if any, ability to inhibit thrombin and a reduced affinity for heparin. Therefore, while there is strong evidence of proteolytic modification of antithrombin by thrombin when they exist in complex formation, how far this complex dissociates *in situ* (i.e. without perturbation of the equilibrium of the initial reaction shown above by preparative techniques involving dilution etc.) to yield free modified antithrombin, prior to irreversible covalent association within the complex, remains to be determined.

7.2.1.2 HEPARIN–ANTITHROMBIN III INTERACTION

7.2.1.2.1 *Binding characteristics*

Three independent groups initially showed that about one third of all heparin molecules in typical commercial preparations account for the major part of the anticoagulant activity of the polysaccharide (Lam *et al.*, 1976; Höök *et al.*, 1976; Andersson *et al.*, 1976). These active molecules have been isolated in several ways. Sucrose density gradient centrifugation of heparin mixed with antithrombin gives rise to the separation of a stable heparin–antithrombin complex (Lam *et al.*, 1976). Affinity chromatography of heparin on antithrombin-substituted Sepharose (Andersson *et*

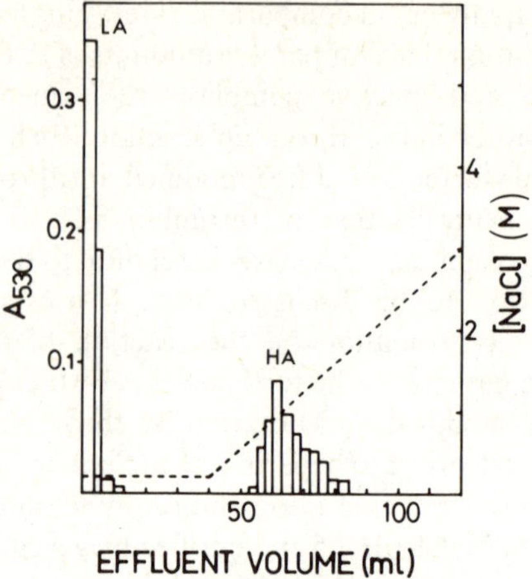

FIGURE 7.3 Chromatography of heparin on antithrombin-Sepharose. A sample (1 mg) of heparin, dissolved in 1 ml of 0.2M NaCl-0.1M Tris-HCl, pH 7.4, was applied to a 3 ml-column of antithrombin-Sepharose, equilibrated with the same buffer at 4°C. After washing with 40 ml of buffered 0.2 M NaCl, the column was eluted with sodium chloride (linear gradient from 0.2 M to 3 M concentration) in 0.1 M Tris-HCl, pH 7.4. Effluent fractions were analyzed for uronic acid by the carbazole reaction (A_{530}), (- - -) NaCl concentration. LA = low-affinity heparin; HA = high-affinity heparin. (From Höök *et al.*, 1976; reproduced with permission of the publishers.)

al., 1976; Höök *et al.*, 1976) is a most convenient procedure, as it allows the separation of large amounts of material. An example of such chromatography is given in Figure 7.3. Two major heparin fractions, one with low-affinity and the other with high-affinity for the immobilized antithrombin are separated in this manner at pH 7.4 and ionic strength 0.14. The high-affinity fraction has an anticoagulant activity in a whole-blood assay of 250–300 B.P. units/mg while the activity of the low-affinity fraction is less than 20 B.P. units/mg. The two fractions also show a large difference in activity when tested in an assay system containing purified antithrombin and thrombin (Höök *et al.*, 1976). The gross chemical structure of the two fractions is very similar. Although this technique yields two major fractions, the high-affinity fraction is heterogeneous with respect to anticoagulant activity (Lindahl *et al.*, 1979). In fact, experiments have shown that high-affinity heparin species eluting at very high ionic strengths during affinity chromatography on immobilized antithrombin have much higher binding constants than heparin high-affinity species eluting early in the chromatogram. Preparations of monodisperse heparin (by gel chromatography) of low molecular weight (6,000) (Jordan *et al.*, 1979) and relatively higher molecular weight (15,000) (Björk *et al.*, 1979) are also heterogeneous with respect to their abilities to bind to and activate antithrombin.

A large body of work has recently been performed concerning the changes induced by unfractionated heparins and high- and low-affinity heparins on the conformation of antithrombin as demonstrated by various spectral techniques including ultraviolet difference spectroscopy, circular dichroism and fluorescence.

Studies in the far ultraviolet CD spectrum of antithrombin in the presence of either unfractionated heparin (Villaneuva and Danishefsky, 1977) or monodisperse heparin ($M = 15{,}300$) of both high- and low-affinity type (Nordenman and Björk, 1978a) have shown that the binding of heparin does not lead to major changes in the secondary structure of the protein. Furthermore, no major change in the size or shape of antithrombin in the presence of unfractionated heparin has been recorded in small angle X-ray diffraction studies (Furugren *et al.*, 1977). In contrast, significant spectral changes of antithrombin have been monitored with unfractionated heparin by solvent perturbation spectra (Villaneuva and Danishefsky, 1977) and for monodisperse high- and low-affinity heparin by near ultraviolet absorption difference spectra (Figure 7.4), circular dichroism (Figure 7.5) (Nordenman and Björk, 1978a) and fluorescence spectra (Figure 7.6) (Nordenman *et al.*, 1978b). In all cases, the highly active, high-affinity heparin causes considerably greater spectral changes than the relatively inactive, low-affinity fraction.

The nature of all spectral changes observed on the addition of unfractionated heparin, or high- or low-affinity heparin to antithrombin suggests that the heparin binding produces local perturbations of the environment of some aromatic residues; chiefly one or more of the tryptophan residues of which there are six residues in antithrombin (Kurachi *et al.*, 1976). Perturbations of tryrosine and phenylalanine residues may also occur.

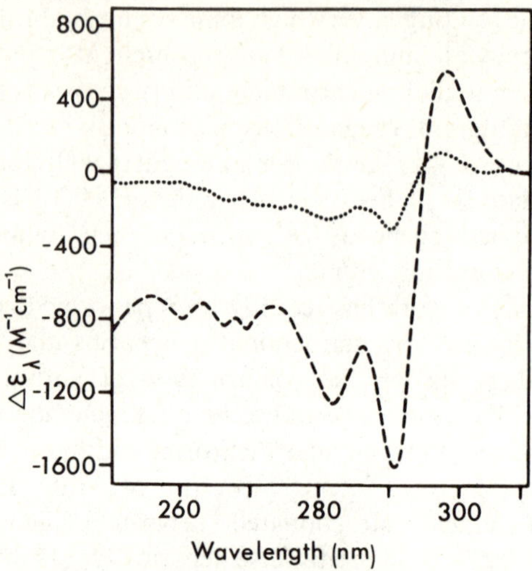

FIGURE 7.4 Ultraviolet absorption difference spectra of human AT due to low-affinity heparin (. . .) or high-affinity heparin (– – –). Protein concentrations were22–26 μM (i.e. 1.2–1.5 g/l), and the solvent was 0.05M Tris buffer, pH 7.4, 0.1M NaCl. The molar ratios between heparin and AT were 4.5 for low-affinity heparin and 2.8 for high-affinity heparin. The unit on the ordinate is the difference between the molar absorptivities of heparin-AT complex and free protein. Measurements were performed at 22°C$\pm$2°C. (Adapted from Nordenman and Björk, 1978a; reproduced with permission of the publishers.)

At the present time, it is difficult to resolve definitely whether these spectral perturbations actually represent a conformational change of antithrombin or local perturbations of solvent environment of the affected residues.

The spectral changes observed have been used to monitor titrations of antithrombin with various heparin fractions to determine binding stoichiometry and equilibrium association constants (Table 7.1). The analyses reveal a stoichiometry of 0.7–1.0 for the binding of both low-affinity and high-affinity heparin to antithrombin. The binding constants measured also reveal that the difference in the affinity of matrix-bound antithrombin for the two major heparin forms is also shown by antithrombin in solution. Furthermore, the ionic strength dependence of the dissociation of

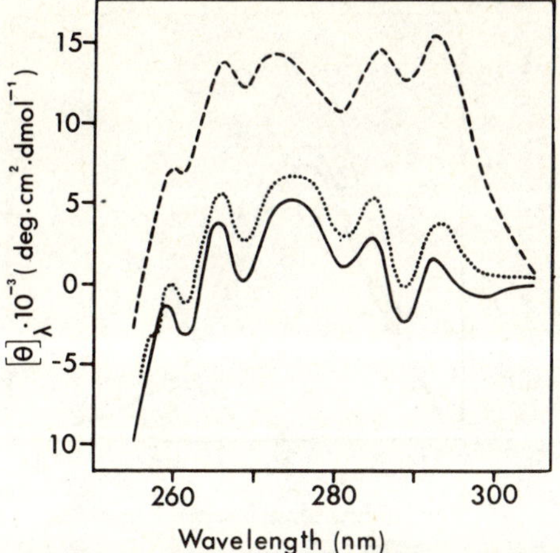

FIGURE 7.5 Near-UV circular dichroism spectra of human AT alone (———) or in the presence of either low-affinity heparin (...) or high-affinity heparin (– – –). Protein concentrations of 21–34 μM (i.e. 1.2–1.9 g/L) were used. The molar ratios between heparin and AT were 4.6 for low-affinity heparin and 1.5 for high-affinity heparin. The solvent was 0.05 M Tris buffer, pH 7.4, 0.1 M NaCl. All measurements were made at $22° \pm 2°$C. No heparin was included in the blanks, as the polysaccharide had no ellipticity in the near-UV region. The unit on the ordinate is molar ellipticity. (Adapted from Nordenman and Björk, 1978a; reproduced with permission of the publishers.)

these complexes showed similar behaviour in both solution and solid phase studies (Nordenman and Björk, 1979).

Careful analysis of the thermodynamics of the interaction of antithrombin with highly active heparin ($M - 6{,}500$) has revealed that this process is an entropy driven event with $\Delta H° = 4.56 \pm 0.09$ kcal/mol and $\Delta S^{37°} = 45.0 \pm 0.16$ entropy units/mol (Jordan *et al.*, 1979).

It is clear that the binding constant, at physiological pH and ionic strength for the highly active, high-affinity heparin fraction is about 1,000 times higher than the binding constant for the relatively inactive, low-affinity fraction. This result strongly supports the contention that tight binding to antithrombin is a prerequisite for heparin anticoagulant activity (Lam *et al.*, 1976; Höök *et al.*, 1976; Andersson *et al.*, 1976). However, it is also important to comprehend the binding of low-affinity heparin in

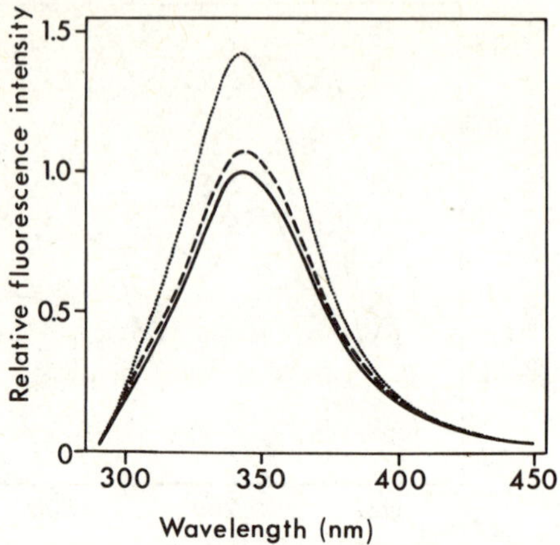

FIGURE 7.6 Corrected fluorescence emission spectra of human anti-thrombin (———) and of its complex with low-affinity heparin (– – –) or high-affinity heparin (. . .). The excitation wavelength was 278 nm. Protein concentrations of 8.3–8.6 μM (i.e. about 0.5 g/L) were used. The molar ratios between low-affinity of high-affinity heparin and antithrombin were 6.3 and 1.6, respectively. The solvent was 0.05M Tris buffer, pH 7.4, 0.1M NaCl. (Adapted from Nordenman *et al.*, 1978; reproduced with permission of the publishers.)

terms of whether it is able to form relatively abortive heparin–antithrombin complexes (for further discussion Section 7.2.1.4.5), which decrease the activity of antithrombin in the anticoagulant reaction. Evidence has been obtained that there may be at least two binding sites on antithrombin for unfractionated heparin (Einarsson and Andersson, 1977) (see Table 7.1). However, it has not been possible to define these sites in terms of a particular heparin fraction.

A clearer demonstration of the competitive nature of binding of low-affinity and high-affinity heparin to the same or different sites on antithrombin has been demonstrated by Danielsson and Björk (1978). Affinity chromatography experiments showed that high-affinity heparin was able to displace radio-labelled low-affinity heparin from matrix-linked antithrombin as well as ^{125}I-labelled antithrombin from matrix-linked low-affinity heparin. Quantitative analyses of the ability of low-affinity heparin to displace radio-labelled high-affinity heparin from antithrombin bound to microcrystalline cellulose were also made and were in good agreement with those obtained by other techniques. Their experiments thus suggest that there is only one heparin binding site on the antithrombin molecule. Both low-affinity and high-affinity heparin bind to this site, although with widely different affinities and with different effects on the conformation and activity of the protein; the difference in affinity must be due solely to inherent differences in the structure of the two heparin fractions. Low-affinity heparin may thus lack some structural features necessary for tight binding, or alternatively may possess some features incompatible with such binding.

7.2.1.2.2 *Disaccharide sequence of antithrombin III – binding-fragment of high-affinity heparin*

In terms of analysing the binding region of heparin to antithrombin more closely. Hopwood *et al.*, (1976) employed a clever technique of digesting heparin bound to antithrombin-linked Sepharose with a bacterial heparinase and then isolating the fragment of heparin which remained bound to the antithrombin. Fragments with high-affinity for antithrombin were derived from high-affinity heparin only. The molecular weight of the active fragments, determined by gel chromatography, was in the range of $4–5 \times 10^3$. These findings suggested that binding of heparin to antithrombin may require a specific sequence of variously substituted sugar residues. A further method for the preparation of the binding site in high yield is by partial deaminative cleavage of the unfractionated polysaccharide with nitrous acid followed by affinity chromatography on immobilized antithrombin (Lindahl *et al.*, 1979). Both methods yield antithrombin-binding heparin fragments of 12 to 16 monosaccharide units. The exact number of disaccharide units involved has not been established. It is important to note that these fragments have low, but finite anticoagulant activity (33 B.P. units/mg) (Laurent *et al.*, 1978). When expressed on a molar basis it is approximately 10% of that of the parent high-affinity heparin. Obviously a heparin chain longer than a dodecasaccharide is required for full activity and probably requires a thrombin-binding site for full activity (Laurent *et al.*, 1978) (Section 7.2.1.4.4).

Compositional analysis of deamination products of heparin species having high-affinity for antithrombin (Rosenberg, 1978; Rosenberg *et al.*, 1978; Rosenberg and Lam, 1979; Lindahl *et al.*, 1979) and studies of periodate oxidation of heparin (Fransson and Lewis, 1979) have demonstrated that non-sulphated L-iduronic acid occurs predominantly in the high-affinity heparin fraction. This residue represents an essential structural feature required for the anticoagulant activity of heparin. Lindahl *et al.* (1979) have shown that this component was concentrated to the antithrombin-binding regions of the high-affinity heparin molecules, amounting to approximately one residue per binding site. Figure 7.7 shows a tentative structure of the possible heparin sequence likely to have a high affinity for antithrombin (Lindahl *et al.*, 1979); it probably represents only a fraction of the total possible antithrombin-binding site structures. Compositional analysis of products formed on deamination of the isolated antithrombin-binding heparin fragments showed significant differences between the materials derived from bovine lung and from pig intestinal mucosa, suggesting that the antithrombin-binding site does not consist of one unique saccharide sequence but has a variable structure. This variability is probably the reason for the wide affinity range displayed by high-affinity molecules (as in Figure 7.3) on affinity chromatography.

Another approach in establishing the critical sequence of disaccharide residues in heparin for antithrombin binding has been used by Rosenberg and coworkers (Rosenberg, 1978; Rosenberg *et al.*, 1978; Rosenberg and Lam, 1979). In this case, in using a starting preparation of monodisperse heparin of low molecular weight they have isolated tetrasaccharide fragments (by nitrous acid deamination) which reflect compositional differences in high-affinity and low-affinity heparins of this preparation. Analysis of these fragments from porcine heparin (Rosenberg, 1978; Rosenberg and Lam, 1979) gives rise to a tetrasaccharide sequence for high-affinity heparin of L-iduronic acid→N-acetylated D-glucosamine-6-sulphate→D-glucuronic acid→N-sulphated D-glucosamine 6-sulphate. This sequence is identical to residues 3–6 in the antithrombin-binding fragment from mucosal heparin obtained by Lindahl *et al.* (1979) (Figure 7.7). Of particular

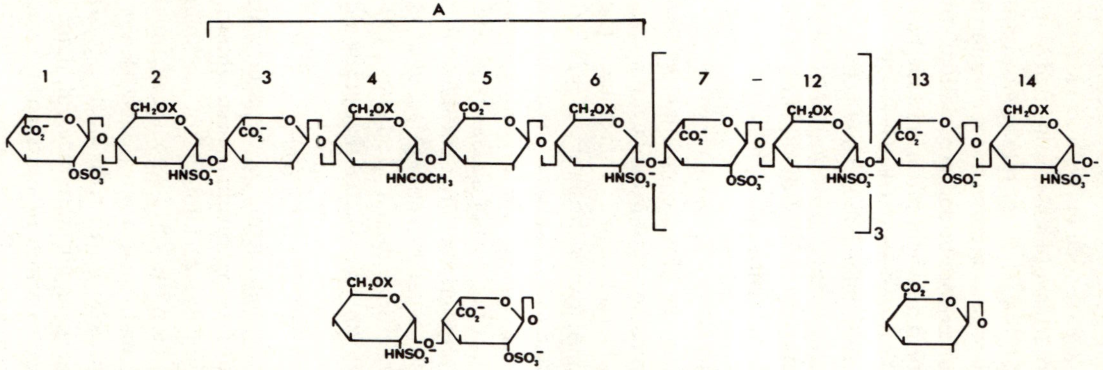

FIGURE 7.7 Tentative structure for the antithrombin-binding site of pig mucosal heparin. X = H or SO_3^- (Lindahl *et al.*, 1979). The total number of disaccharide units per antithrombin-binding site has been set at seven but may instead be six or eight, such deviations from the structure shown should presumably affect only the portion of the sequence to the right of position 7. The tetrasaccharide sequence in region A has been obtained by Rosenberg *et al.* (see text). Lindahl *et al.* (1979) maintain that sugar units at positions 4, 5 and 13 represent structural variants compatible with (but not required for) anticoagulant activity. There is general agreement that the non-sulphated iduronic acid residue at position 3 is required for high-affinity binding to antithrombin. Recent work by Riesenfeld *et al.* (1979) would suggest that the minimal binding fragment probably only involves 10 sugar units, which means that the sugar units 11–14 and its variant are irrelevant (see Note in Added Proof p248).

interest is the corresponding tetrasaccharide unit from low-affinity heparin (anticoagulant activity, 1Z USP units/mg), which contained a sulphated iduronic acid residue in the iduronic position.

It is clear, then, that the occurrence of a non-sulphated iduronic acid residue in heparin, which is a relatively rare event, confers a high probability that it will present a correct sequence for high affinity to antithrombin and therefore be endowed with relatively potent anticoagulant activity. The origin of this residue has not been established. It may occur through a particular mechanism of biosynthesis or may result from limited sulphatase activity on a molecule once formed. Studies in this direction will be of great interest in terms of the biological control of the anticoagulant activity of heparin (see Note in Added Proof, p 248).

7.2.1.3 HEPARIN–THROMBIN INTERACTION

Thrombin does not exist in the blood circulation *per se* but, when it is formed from its precursor, prothrombin, it may exist in two major active forms, α-thrombin and β-thrombin. The β-thrombin ($M = 28{,}000$) is formed from α-thrombin ($M = 39{,}000$) by degradation of the A-chain and excision of a small carbohydrate-containing fragment. The activities of the α- and β-forms of thrombin are nearly equivalent in terms of amidase and esterase activities, whereas their clotting activities are quite different, with the α-thrombin being considerably more active (Machovich *et al.*, 1975a; Machovich *et al.*, 1976; Borsodi and Machovich, 1979).

Nordenman and Björk (1978b) have shown that heparin does not bind to prothrombin, so that the heparin-binding site of thrombin must be unmasked or created following activation of prothrombin. The preferential interaction of heparin with α-thrombin as compared to β-thrombin has been shown by Machovich and coworkers by a number of techniques. Unfractionated heparin has been shown to stabilize α-thrombin, but not β-thrombin, to thermal inactivation (Machovich *et al.*, 1976; Machovich and Arányi, 1978). Further, the inactivation rates of α- and β-thrombin by antithrombin were different, namely α-thrombin was more sensitive to antithrombin than β-thrombin (Machovich *et al.*, 1976; Orakzai and Machovich, 1977). It therefore appears that the heparin-sensitive form of thrombin is α-thrombin and the insensitive form β-thrombin.

Apart from these studies, the interaction of unfractionated heparin with thrombin has been directly shown by precipitation of thrombin by heparin in 0.1M NaCl, 0.02M Tris HCl, pH 7.5 (Owen, 1977) and isolation of a thrombin–heparin complex on Sephadex G-200 in 0.05M Tris HCl at 0°C (Machovich *et al.*, 1975c). In addition, a number of studies have shown that unfractionated heparin linked to Sepharose is able to bind thrombin (Gentry and Alexander, 1973; Machovich *et al.*, 1975c; Danishefsky *et al.*, 1976; Hatton and Regoeczi, 1977). In terms of strength of binding, as measured by the ionic strength of solvent required to elute thrombin from the column, highly variable results have been obtained ranging from $I = 0.5$ (Machovich *et al.*, 1975c) to considerably lower ionic strengths (Hatton and Regoeczi, 1977; Griffith *et al.*, 1978). The reasons for these variations are not clearly understood. However, the binding property has been used successfully for the purification of highly purified α-thrombin (Nordenman and Björk, 1977). Studies with thrombin-linked Sepharose columns have shown that both high-affinity heparin and low-affinity heparin elute at the same ionic strength range operative at 0.5M NaCl (Nordenman and Björk, 1978b) or 0.25M NaCl (Jordan *et al.*, 1979). A range of binding constants for heparin–thrombin interaction may operate within a heparin preparation (Nordenman and Björk, 1978b). Nordenman and Björk (1978b) have also shown that low-affinity heparin-linked and high-affinity heparin-linked Sepharose columns have the same affinity for thrombin. Similar results have been obtained utilizing a crossed immunoelectrophoresis technique (Holmer *et al.*, 1979). These studies do suggest that all heparin molecules in typical unfractionated heparin preparations bind to thrombin with similar affinity. In contrast, however, is a recent study by Griffith *et al.* (1978) in which commercial heparin (135 USP units/mg) was fractionated by human α-thrombin-linked agarose affinity chromatography. An unbound fraction was washed from the column with 0.01M Tris HCl and the bound heparin could also be eluted at low ionic strength of 0.025M NaCl. The bound heparin was highly active anticoagulant (500 USP units/mg). In view of the vastly different results obtained in this study as compared to the bulk of other evidence, these studies must be confirmed to demonstrate a highly active heparin–thrombin species, which may exist under physiological conditions.

At present there has been no satisfactory experimental technique to evaluate a dissociation constant for the thrombin–heparin interaction. Studies so far published suggest that heparin does not induce any significant spectroscopic changes in thrombin, as judged by ultraviolet difference spectra and solvent perturbation spectral studies (Villaneuva and Danishefsky, 1979). However, in a preliminary communication by Rosenberg *et al.* (1980), the thrombin binding to heparin has been measured by polarization fluorescence. This method reveals that thrombin exhibits two binding sites for highly active or relatively inactive heparins of molecular weight 6500 with association constants of $1.25 \times 10^6\ M^{-1}$ and $1.25 \times 10^6\ M^{-1}$ respectively.

A number of studies, utilizing kinetic analysis of the heparin effects on thrombin activity alone, have invoked a purported thrombin–heparin interaction. The heparin effect on the hydrolysis of synthetic substrates by thrombin, namely Bz Phe Val Arg Na N (S2160) and Tos Gly Pro Arg Na N (Smith, 1977; Hatton and Regoeczi, 1977; Griffith *et al.*, 1979) has been used to establish an interaction of unfractionated heparin with thrombin. Unfortunately, conflicting results have been obtained, as heparin has been shown to have an inhibitory effect (Smith, 1977; Li *et al.*, 1974; Griffith *et al.*, 1979) and an enhancing one (Hatton and Regoeczi, 1977; Griffith *et al.*, 1979) or no effect at all (Nordenman and Björk, 1978b; Jordan *et al.*, 1979). Nordenman and Björk (1978b) (see also Bartl, 1978) claim that the inactivation of thrombin by heparin, which had been measured by use of synthetic substrate S2160, was caused by direct interaction between heparin and substrate S2160 and that, in fact, thrombin activity is not affected by heparin. It is difficult at this stage to draw any definite conclusions from this type of experiment.

7.2.1.4 ANTITHROMBIN–THROMBIN–HEPARIN INTERACTION

7.2.1.4.1 *Gross correlations of chemical features of heparin with anticoagulant activity*

Uronic acid content Taylor *et al.* (1973) have demonstrated that the L-iduronate content of heparin preparations obtained from different sources varied from 50 to 90% of the total uronic acid content and that variation in anticoagulant activity of these preparations was not pronounced (Table 2.1). However, this assessment did not take into account the possible role of particular disaccharide sequences present in these heparins. A feature of their comparative results was the apparent discontinuity of anticoagulant activity in relation to the chemical composition of samples varying from heparan sulphate to heparin. The discontinuous nature of anticoagulant activity is also evident on fractions obtained on ion exchange Dowex 3M. Fractions eluted at [NaCl] >1.5M had high anticoagulant activity (Cifonelli, 1974) and probably reflects the requirement of heparin molecules with high charge density for optimal anticoagulant activity.

Fransson and Lewis (1979) have demonstrated that, when heparin is subjected to periodate oxidation at pH 3.0 and 4°C (conditions which selectively oxidize D-glucuronic acid associated with N-acetylglucosamine units (Fransson, 1978)), negligible reduction of anticoagulant activity occurs. On the other hand, when oxidation is performed at pH 7.0 and 37°C (all non-sulphated uronic acids are oxidized), the anticoagulant activity is completely abolished (see also Braswell, 1968). It was also noted that there were no periodate-resistant D-glucuronic acid residues in the heparin preparation (i.e. D-glucuronic acid associated with N-sulphated glucosamine may be resistant to oxidation at pH 3.0 and 4°C (see Section 4.1.4)). Therefore, it may be supposed that non-sulphated L-iduronic acid residues are critical structural elements of an 'active site' in heparin. It is also of interest that Lindahl *et al.* (1979) have shown that the antithrombin-binding heparin fragment, obtained by deaminative cleavage of parent heparin, when subjected to periodate oxidation (pH 7.0, 37°C) lost all of its anticoagulant activity. These results are in accord with the importance of non-sulphated iduronic acid residue in the antithrombin binding fragment of high-affinity heparin (Section 7.2.1.2.2.).

Some information as to the relative contributions of the carboxyl groups of the uronic acid residues of heparin has been gained from their chemical modification. Among the derivatives that have been examined are heparin methyl ester, heparinyl glycine and its methyl ester (Danishefsky *et al.*, 1977; Danishefsky, 1977). The conclusions are that free carboxyl groups are critical for activity of fibrinogen clotting times either with pre-incubation with antithrombin and fibrinogen or pre-incubation with antithrombin and thrombin, although conclusions regarding specific contributions of carboxyl groups are complicated by the potential steric hindrance effects of the new groups in the derivatives. It should be emphasized that none of the derivatives are as active as heparin in any of the assay systems.

Sulphate content The influence of ester sulphate content on the anticoagulant activity has been difficult to assess, because methods have not been available to specifically remove these groups under suitably mild conditions. However, variations of molar ratios of ester sulphate ratios to hexosamine from approximately 0.9 to 1.6 have been found in heparins having reasonably high anticoagulant activities (Taylor *et al.*, 1973) (Table 2.1). Kosakai *et al.* (1978) have shown that a portion of 6-O-sulphates in glucosamine residues does not contribute significantly to anticoagulant activity of heparin.

The N-sulphate groups appear to be important for activity. Complete loss results in removal of anticoagulant activity (e.g. Danishefsky *et al.*, 1977). Partial release of N-sulphate groups (7–30%) by acid hydrolysis (Barlow *et al.*, 1961; Cifonelli, 1974) or by solvolytic N-desulphation of the pyridinium salt of heparin in dimethyl sulphoxide (Nagasawa *et al.*, 1977) have exhibited mild losses of activity but in varying proportions. Cifonelli (1974) has reported that a commercial hog mucosal heparin sample having 9% of hexosamine units N-desulphated showed a loss of over 30% of the anticoagulant activity. Riesenfeld *et al.* (1977) have obtained hydrolyzed samples having up to one-third of the amino groups of their glucosamine residue unsubstituted, but they still retained appreciable activity. Nagasawa *et al.* (1977) have shown that the anticoagulant activities of various heparin preparations prepared by N-acetylation of partially N-desulphated heparins were nearly equal to those of the corresponding starting N-desulphated heparins. Notably, N-resulphation restored the anticoagulant activity to 94% of that of the original heparin (see also Riesenfeld *et al.*, 1977). Furthermore, a preparation with a degree of N-desulphation of 42% retained 62% of the original activity. Their results show that, if a change in the chemical structure of heparin is restricted to N-desulphation, the loss of biological activity is slow and stepwise and not rapid and definitive as reported previously (Foster *et al.*, 1963; Stivala *et al.*, 1967; Yosizawa *et al.*, 1967).

A relationship of N-sulphate group distribution along the heparin chain to anticoagulant activity was obtained by treating fractions with butyl nitrite and subsequent analysis of oligosaccharides by gel chromatography (Cifonelli, 1974). It appears that, whereas sequences of single repeating N-acetyl glucosamine residues separated by adjacent units of N-sulphated glucosamine residues do not affect the anticoagulant activity of heparin as noted by the occurrence of such sequences in whale heparin (Cifonelli and King, 1972), the presence of block N-acetyl glucosamine residue sequences decreases the anticoagulant activity. This is in accord with the presence of these block structures in by-products of heparin, which have low anticoagulant activity.

Other glycosaminoglycans No other naturally occurring glycosaminoglycan can mimic the effects of heparin in terms of its anticoagulant action. Teien *et al.* (1976) have shown that heparin, human aorta heparan sulphate and porcine intestinal mucosa dermatan sulphate exhibit anticoagulant activity as measured by the APTT test. About 70 times higher concentrations of heparan sulphate and dermatan sulphate were required to obtain the equivalent effect of heparin. Other glycosaminoglycans such as the chondroitin sulphates, keratan sulphate and hyaluronate have no effect. Similar conclusions were obtained from studies on the amidolytic activity of thrombin, although differences were monitored for the non-heparin-like glycosaminoglycans. It remains to be determined whether the effects registered with heparan sulphate and dermatan sulphate are *bona fide* properties of these molecules or due to heparin impurities.

7.2.1.4.2 *Ionic strength and pH dependence*

Early studies by Lasker and Stivala (1967) demonstrated that the anticoagulant activity of unfractionated heparin is not sensitive to salt concentrations up to 0.5M NaCl. At higher salt concentrations a large decrease in activity occurs between 0.5 and 1.0M NaCl. Machovich *et al.* (1975b) found that the rate of thrombin inactivation by antithrombin and heparin was maximal at an ionic strength between 0.1 and 0.2. For NaCl or KCl, the presence of $CaCl_2$ (1mM) was without effect on the reaction (Machovich, 1976).

Nordenman and Björk (1979) have studied the interaction of high- and low-affinity heparins with antithrombin by dissociation at increasing concentrations of NaCl as measured by near-UV circular dichroism, tryptophyl fluorescence of the protein and by direct titration. The mid-points of dissociation curves for the system heparin–Sepharose and antithrombin in solution were 0.15 and 0.85M NaCl for low- and high-affinity heparin respectively. This is in approximate agreement with the elution behaviour of the two fractions on antithrombin–Sepharose. These values are dependent on the protein concentration. It is of interest that the degree of low-affinity heparin binding to antithrombin in the range of 0.02–0.4M NaCl increased with decreasing ionic strength, whereas the high-affinity heparin–antithrombin interaction did not vary in this range of ionic strength. The interaction of high-affinity heparin and bovine antithrombin as measured by fluorescence in 0.14M NaCl showed a mid-point for dissociation at a $CaCl_2$ concentration of 0.3M.

The binding of high-affinity heparin to antithrombin is only preserved between pH values of about 5 and 9 (Nordenman and Björk, 1979). The decrease in the interaction of heparin and antithrombin at pH values > 8.0 has been interpreted as being due, in part, to the decrease in charge–charge interactions that operate (possibly via lysyl residues of antithrombin) between the two molecules (Jordan *et al.*, 1979) (see also Figure 7.8).

7.2.1.4.3 *Protein binding fragments*

An assessment of the important amino acid residues of thrombin and antithrombin involved in binding to heparin has been made

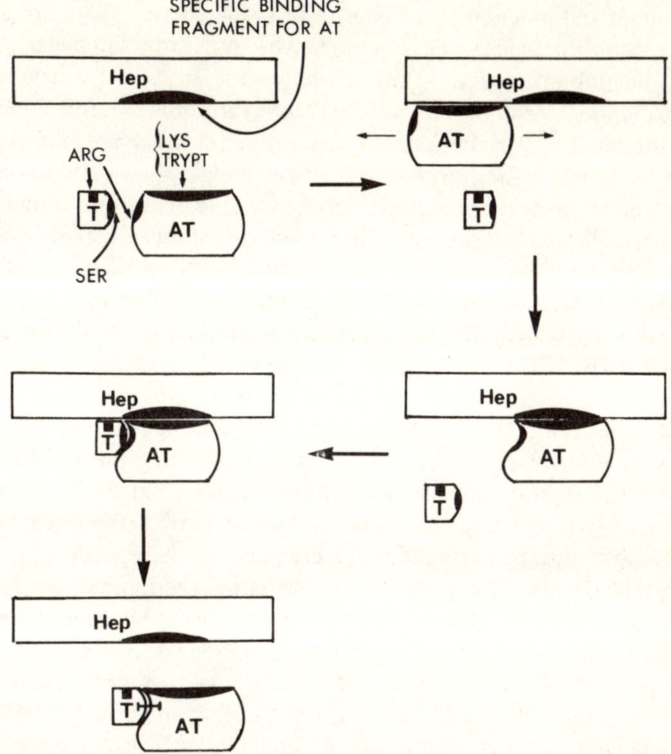

FIGURE 7.8 A model of the thrombin-heparin-antithrombin III interaction in which the contribution of various amino acid residues is depicted (see text for details).

essentially through the effect of their specific chemical modification. Normally, these effects cannot distinguish modification of an amino acid at the binding site, or one distant from it which may induce a conformational change in the binding site. Furthermore, the addition of bulky side groups to the amino acids may offer steric hindrance to binding. Notwithstanding these problems, it is generally viewed that these chemical modifications are, in fact, affecting the active binding sites. The various modifications that have been used and their effect on the thrombin–antithrombin reaction and the binding of various components have been summarized in Table 7.2. These results suggest that there are at least four binding regions of importance in the thrombin–heparin–antithrombin reaction; 1) the serine residue at the active site of thrombin, 2) the arginine residue(s) at the thrombin reactive site of antithrombin, 3) arginine residues in thrombin which are required for heparin binding and 4) lysine and tryptophan residues in antithrombin that are required for the binding of antithrombin to heparin. Some caution should attend the effects of modifying tryptophan residues in antithrombin with N-bromosuccinamide, as peptide scission may occur with this particular treatment (Björk *et al.*, 1979). A model depicting these interactions is shown in Figure 7.8.

7.2.1.4.4 *Molecular weight dependence of anticoagulant activity*

When polydisperse heparin polysaccharide chain preparations have been fractionated according to molecular weight, it has been invariably found that the anticoagulant activity of the fractions increases with the degree of polymerization of the heparin chains. This has been clearly shown for heparin fractionated by gel chromatography (Table 7.3) (Johnson and Mulloy, 1976; MacGregor *et al.*, 1978; Laurent *et al.*, 1978; Lane *et al.*, 1978), by ethanol solubility (Liberti and Stivala, 1967), on ion-exchange resins including Ecteola (Laurent, 1961; Lasker and Stivala, 1966), on Dowex 1 (Di Ferrante *et al.*, 1970), DEAE-Sephadex and protamine-Sepharose (Piepkorn *et al.*, 1978) and by high pressure liquid chromatography (Rodriquez and Vanderwielen, 1979). In spite of the marked fractionation of heparin in terms of its anticoagulant activity by these various techniques, there has been no substantial corresponding fractionation on the basis of chemical composition. There is also little correlation between mean molecular weight of various unfractionated animal heparin preparations and anticoagulant activity (Barlow *et al.*, 1961), which may reflect different molecular weight distributions (Barrowcliffe *et al.*, 1977).

A quantitative relationship between molecular weight and anticoagulant activity, in relation to heparin binding to antithrombin, was confirmed by the demonstration of the enrichment of high molecular weight material with high-affinity on

TABLE 7.3
Analytical Data for Fractions Obtained by Gel Chromatography of Heparin

Fraction No.	Percentage of total material	Molecular weight	K_{av} on Sephadex G-100	Anti-coagulant activity (units/mg)
1	0.8	36,000	0.05	139
2	3.8	27,000	0.09	188
3	7.6	25,000	0.14	184
4	12.4	19,200	0.21	179
5	16.3	16,500	0.24	181
6	19.1	13,000	0.27	159
7	17.4	10,900	0.35	119
8	13.0	7,800	0.47	109
9	6.9	5,600	0.59	4
10	2.7	—	0.68	—
	100.0	$M_w = 14{,}200$ $M_n = 11{,}200$		

K_{av} values were based on peak positions obtained in analytical gel chromatograms on Sephadex G-100.-, insufficient material for analysis. The molecular-weight averages (M_w and M_n) are for the unfractionated material. (From Laurent *et al.*, 1978; reproduced with permission of the publishers.)

antithrombin–Sepharose columns and relatively lower molecular weights for material with low-affinity for the antithrombin columns (Laurent *et al.*, 1978).

In view of the finding that the active binding region of heparin has a molecular weight of 4,000 (i.e. approximately corresponding to that of a dodecasaccharide (six disaccharides)), it was suggested by Laurent *et al.* (1978) that the increase with molecular weight in overall anticoagulant activity (when measured by the BP method, APTT or thrombin inhibition) could be related to the probability of finding this binding site in a preparation. They showed that the relative amount of high-affinity material in heparin fractions of different molecular weight followed a probability function. There was much less high-affinity heparin in the low molecular weight region than at higher molecular weights. This was, however, not the full explanation for the molecular weight dependence since the activity of pure high-affinity heparin was also molecular weight

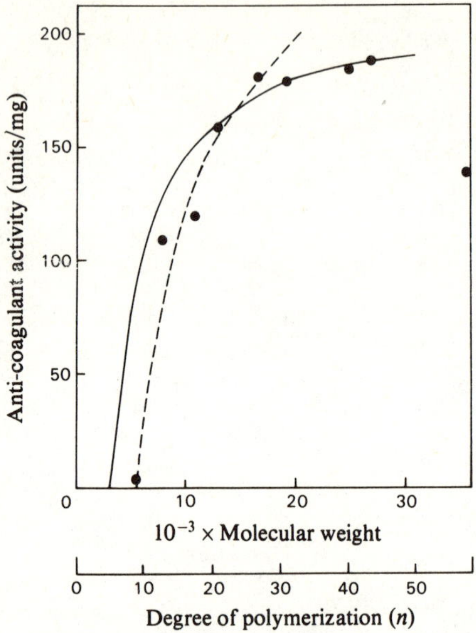

FIGURE 7.9a Anticoagulant activity of heparin fractions (Table 7.3) plotted versus molecular weight and degree of polymerization. The continuous line follows the relationship: activity $\propto (n-5)/n$ and has been adjusted to the value at the molecular weight 16,500. The broken line follows the relationship: activity $\propto (n-9)/n$ and has been adjusted to the experimental points below 13,000. (Adapted from Laurent *et al.*, 1978; reproduced with permission of the publishers.)

dependent. Furthermore, there is an agreement obtained by plotting the anticoagulant activity of heparin versus molecular weight with probability best described by the function, activity = constant. $[(n-9)/n]$, where n is the number of disaccharides in a heparin molecule (Figure 7.9a). This demonstrates that a heparin sequence of ten rather than six disaccharides is required for activity. Only at higher range of molecular weights does the relationship drastically break down, most likely due to the presence of polysaccharide chains linked to a peptide core as a remnant of endoglycosidase action (fragments containing the polysaccharide–protein linkage region are expected to behave differently from other heparin fragments since they have a different chemical composition and, in addition, a branched structure).

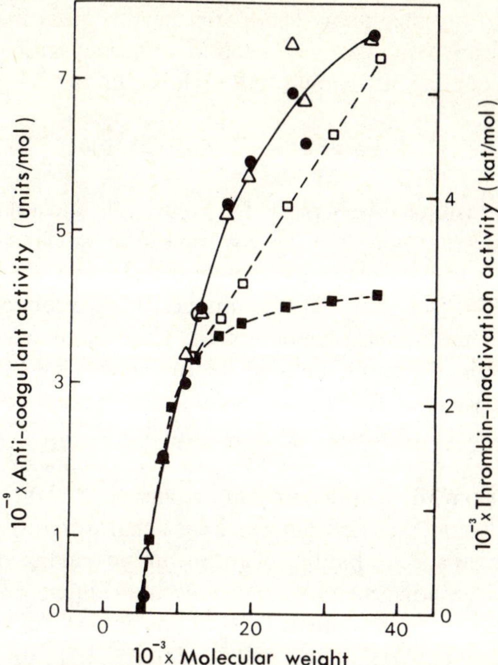

FIGURE 7.9b Anticoagulant activity (●) and the thrombin-inactivating activity (△) of HA-heparin as a function of molecular weight. The activities are given in units per mol. The scales have been adjusted so that the curves should overlap. Two theoretical curves have also been incorporated. One has been drawn according to equation 7.1 and adjusted to the experimental points in the low-molecular-weight region (■). It shows the probability of finding a thrombin-binding site adjacent to a single antithrombin-binding site on each heparin molecule. The second curve (□) has been obtained by multiplying the first curve with the probable number of antithrombin-binding sites per heparin molecule. (kat = katal = activity corresponding to conversion of 1 mol of synthetic chromogenic substrate per sec.) (From Laurent *et al.*, 1978; reproduced with permission of the publishers.)

A heparin molecule then may contain a dodecasaccharide sequence required for complete antithrombin binding but this is not enough to confer full anticoagulant activity on the molecule. An additional octasaccharide (4 disaccharides) is required for anticoagulant activity. On this basis, Laurent *et al.* (1978) have proposed that a thrombin molecule binds to this octasaccharide, adjacent to antithrombin. Therefore, the thrombin inactivating activity (ζ) of a high-affinity heparin molecule with one antithrombin binding site should follow the equation

$$\zeta = \text{const.}[(n-9)/(n-5) \qquad 7.1$$

This relationship has been plotted in Figure 7.9b and adjusted to fit the experimental points in the low molecular weight range. The discrepancy at high molecular weight of heparin is decreased by invoking the fact that there is an increasing probability with increase in molecular weight that the chain contains a second antithrombin–thrombin binding site. In fact this has been shown by Rosenberg *et al.* (1979), who isolated two monodisperse high-affinity heparin fractions; one of molecular weight 20,000 with an anticoagulant activity of 738 USP units/mg, and one of molecular weight 7,000 with an anticoagulant activity of 363 USP units/mg.

In conclusion, this model in which the heparin acts as a catalytic surface for the specific binding of antithrombin and the less specific binding of an adjacent thrombin molecule (Figure 7.8) satisfactorily predicts, on a statistical basis, the molecular weight dependence of heparin anticoagulant activity. It is to be noted, however, that other mechanisms may be required to explain the finite anticoagulant activity of the antithrombin–heparin binding fragment (providing that it is relatively monodisperse) and that of low-affinity heparin.

The anticoagulant activity of macromolecular heparin and its relationship to high- and low-affinity heparin chains has been subject to limited study only. Yurt *et al.* (1977) have demonstrated that macromolecular heparin isolated from rat peritoneal mast cells has very low anticoagulant activity, whereas macromolecular heparin from rat skin has reportedly significant activity of 123 USP units/mg (Horner, 1971). The relationship between synthesis of high- and low-affinity heparin chain segments within macromolecular heparin and/or the possibility of active chain formation by limited sulphatase activity (Section 7.2.1.2.2) should provide valuable information as to the control of such anticoagulant activity *in vivo*. Studies in this direction have recently been performed by Horner and Young (1979). Heparin proteoglycan from rat skin was fractionated into no-affinity, low-affinity and high-affinity forms for human antithrombin-linked Sepharose. The proteoglycans were then depolymerized by platelet endoglycosidase. The content of no-affinity, low-affinity and high-affinity forms of heparin chain fragments for high-affinity and low-affinity proteoglycan were in the proportions of 45:33:22 and 54:39:7 respectively. These experiments suggest that heparin chains within a proteoglycan molecule exhibit full range affinity to antithrombin. Furthermore, it was demonstrated that the anticoagulant activity of the proteoglycan is the sum of the anticoagulant activities of its constituent heparin chains. It was concluded therefore that the anticoagulant activity of heparin is not generated by depolymerization of the multi-chain form.

7.2.1.4.5 *Mechanism of interaction*

There have been essentially three basic mechanisms proposed for optimal effect of heparin-mediated-thrombin neutralization by antithrombin. Firstly, the interaction of heparin with antithrombin has been proposed to be the primary event of the heparin effect in the thrombin–antithrombin reaction (Rosenberg and Damus, 1973). As considered in the previous sections (Sections 7.2.1.2.1 and 7.2.1.2.2) only a small proportion of heparin molecules within commercial heparin preparations may avidly bind to antithrombin and thereby induce a reversible conformational change in the protein so that it may act as a more effective inhibitor of thrombin. This view is supported by the fact that effective inhibition of thrombin by antithrombin would only occur for antithrombin binding to a particular sequence of disaccharides within a heparin molecule. Although there appears to be no specific sequence on the heparin molecule required for thrombin binding, the direct interaction between these two molecules would deem it necessary to account for this interaction in terms of thrombin inactivation. This has led to two further hypotheses. It has been suggested that heparin binding to thrombin makes the enzyme more suitable (as yet in an unknown fashion) to neutralization by antithrombin (Gitel, 1975; Machovich *et al.*, 1975c). However, there is little evidence to support this mechanism as singularly determining the heparin effect on thrombin neutralization. Indeed, as a physiological reaction it is a tenuous one; thrombin is not present in normal plasma, so that it would require heparin transfer from heparin carrier proteins to thrombin as it is produced. As pointed out by Barrowcliffe *et al.* (1978), if heparin activates thrombin for a substrate competitive inhibitor such as antithrombin then it would be expected to activate the enzyme also for its normal substrate. The opposite seems to be the case, as heparin alone appears to inhibit the action of thrombin on fibrinogen (Lages and Stivala, 1973).

The most appealing mechanism to account for the essential features of heparin interaction with thrombin and antithrombin is that elaborated by Laurent *et al.* (1978) for the molecular weight dependence of heparin anticoagulant activity (Section 7.2.1.4.4). In this mechanism, optimal reaction occurs when antithrombin interacts with a specific disaccharide sequence in the heparin molecule and thereby activating the inhibitor for action on a thrombin molecule that resides adjacent to it, bound to the same heparin molecule (Figure 7.8). This mechanism is also supported by 1) protein chemical modification studies (Table 7.2) in which thrombin- and antithrombin-binding sites for heparin may be independently modified without affecting the thrombin–antithrombin reaction, 2) the binding of heparin to antithrombin–thrombin complex (Carlström *et al.*, 1977) and 3) the ability of heparin, bound to antithrombin-linked Sephadex, to bind thrombin (Pomerantz and Owen, 1978).

In the following discussion, the evidence on the heparin effect on the thrombin–antithrombin reaction will be considered in the light of the above mechanisms and will be finally discussed in terms of an overall mechanism.

Catalytic effect of heparin The quantities of heparin required to give an accelerating effect on the thrombin–antithrombin reaction are remarkably small and are also non-stoichiometric. It was early shown that a rapid rate of thrombin neutralization can be demonstrated with an antithrombin–heparin molar ratio >150 (Biggs *et al.*, 1970), which suggested a catalytic effect of heparin. The catalytic role for heparin in the inhibition of thrombin by antithrombin has been confirmed in recent studies (Björk and

Nordenman, 1976; Kowalski and Finlay, 1979; Jordan *et al.*, 1979). Björk and Nordenman (1976) have shown that one molecule of heparin was able to promote the binding of 30–40 molecules of antithrombin to thrombin *in vitro*, when the reaction was allowed to proceed for a sufficient period of time.

The catalytic action of heparin requires that it is released for further binding once the thrombin–antithrombin complex was formed (Gitel, 1975). This release reaction has been shown by a number of studies. Carlström *et al.* (1977) have demonstrated that the binding of heparin to the antithrombin–thrombin complex was lower than heparin binding to antithrombin alone as judged by ionic strengths of elution on heparin–Sepharose columns, gel electrophoresis and gel chromatography. This has been further shown by a two dimensional immunoelectrophoretic method (Andersson *et al.*, 1977). Jordan *et al.* (1979) using gel chromatography showed that thrombin is capable of releasing [^{14}C]-heparin from [^{14}C] heparin–antithrombin complex and that maximum heparin release occurs at a thrombin–antithrombin molar ratio of 1.02. The binding of heparin to thrombin–antithrombin complex in solution is 100-fold weaker than the interaction of heparin with antithrombin alone (Jordan *et al.*, 1979).

Factors contributing to sub-optimal interactions There is now accumulating evidence in the literature which would indicate that there are several factors which contribute to sub-optimal interactions in thrombin–heparin–antithrombin ternary complex formation. Firstly, abortive binary complexes (as compared to active complexes for optimal reaction) between heparin: antithrombin or between heparin:thrombin may possibly occur under conditions where there is an excess concentration of one of the reactants. Secondly, thrombin proteolysis of antithrombin to yield inactive modified antithrombin (Section 7.2.1.1) should also be considered in this context.

Early indications of these phenomena occurring came from several studies in which, with sufficiently high concentrations of unfractionased heparin, the relative extent or apparent stoichiometry of inhibition of thrombin by antithrombin is decreased without a loss of the rate of the enhancement of the inhibition

(Rosenberg and Damus, 1973; see also Björk and Nordenman, 1976; Marciniak, 1975). This effect was shown to be critically dependent on the thrombin–antithrombin molar ratio (Marciniak, 1975). For excess thrombin, substrate inhibition was evident (see also Yue *et al.*, 1977) which could be relieved by further addition of antithrombin (Marciniak, 1975). Conversely, in the presence of excess antithrombin, the residual activity of antithrombin (as measured by factor Xa assay), when thrombin activity had approached zero, indicated that the utilization of the inhibitor was more extensive in the presence than in the absence of heparin (Marciniak, 1977). A re-examination of these effects has been performed by Owen (1977) who used both unfractionated heparin and a heparin fraction of narrow molecular weight range obtained by fractionation by gel chromatography. The effects of both crude and monodisperse heparin on the inhibition of thrombin by antithrombin are shown in Figure 7.10. Figure 7.10A and B show inhibition of thrombin under conditions of excess antithrombin and thrombin respectively. In both cases, but shown more clearly with excess thrombin (Figure 7.10B), increasing concentrations of heparin caused a decreasing extent of inhibition. With excess thrombin, the level of residual thrombin did not change with up to 120 minutes of incubation. When the heparin preparation obtained by gel chromatography fractionation was used in place of the unfractionated product, the effect on the extent of inhibition was diminished markedly (Figure 7.10C). In that this latter result is unusual, a possible mechanism by which the unfractionated heparin could reduce the extent of inhibition of thrombin by antithrombin would be for a component in the crude heparin to form an unreactive complex with thrombin, or antithrombin or both. Owen (1977) demonstrated by SDS-gel electrophoresis that no apparent proteolysis of unreacted residual antithrombin by thrombin had occurred. While it has been recently established that proteolysis of antithrombin by thrombin does occur in the formation of the inactive complex (detected only under reducing/denaturing conditions) (Fish and Björk, 1979), it remains to be determined whether antithrombin modification occurs without significant concomitant thrombin neutralization.

Further evidence of heparin-mediated abortive complexes has

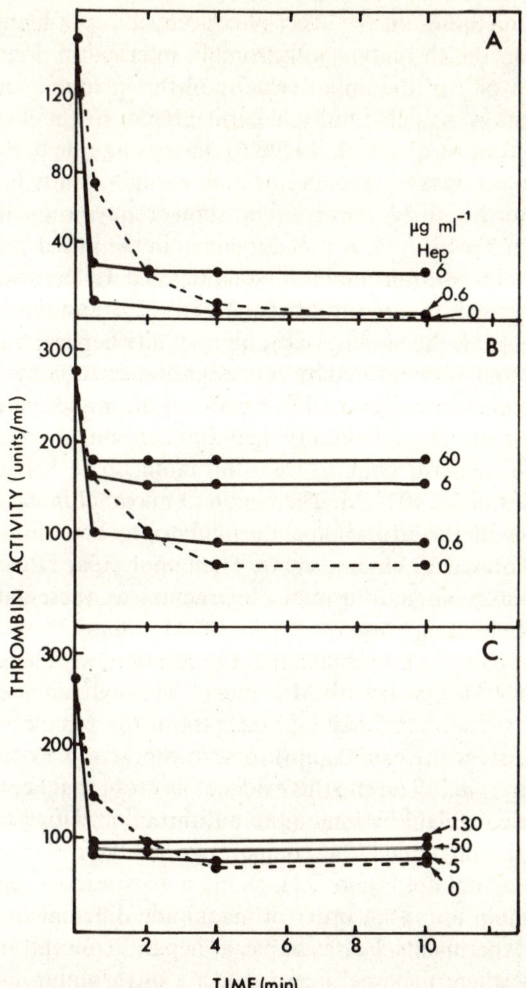

FIGURE 7.10 Effect of heparin on the extent of the thrombin-antithrombin III reaction. To mixtures containing antithrombin III (130 μg/ml final concentration) were added heparin to the (final) concentration indicated and thrombin to a final concentration of either 50 μg/ml (A, inhibitor excess) or 90 μg/ml (B, C, enzyme excess). A: excess inhibitor, unfractionated heparin; B: excess thrombin, unfractionated heparin; C: excess thrombin, heparin fraction obtained by gel chromatography. Dotted lines show control reactions (no heparin). (From Owen, 1977; reproduced with permission of publishers.)

come from studies on the effect of concentration of high-affinity heparin on the thrombin–antithrombin interaction. Figure 7.11 presents a plot of the initial velocity of this reaction versus the concentration of high-affinity heparin present in the incubation mixture (from Jordan *et al.*, 1979). An increasing rate of thrombin inhibition is noted as the concentration of high-affinity heparin is raised from 5×10^{-8} M; the concentration of enzyme and inhibitor were set at 1×10^{-8} M. A pseudoplateau in the initial velocity of the thrombin–antithrombin interaction occurs between high-affinity heparin of concentrations 2×10^{-7} M and 5×10^{-6} M. Also depicted is the binding of the high-affinity heparin fraction to antithrombin as monitored by fluorescence spectroscopy. There is an apparent close correspondence between the initial velocity plot as a function of high-affinity heparin concentration and the amount of heparin bound to antithrombin up to heparin concentrations of 5×10^{-6} M. The region of maximal initial velocity coincides with the saturation of the inhibitor by heparin as judged by the fluorescence binding curve. The bimolecular rate constant for the thrombin–antithrombin interaction at these saturating levels of high-affinity heparin is 8.0×10^{8} M^{-1} min^{-1} as compared to the bimolecular rate constant of the reaction, with heparin not present, of $4.34(\pm 0.44) \times 10^{5}$ M^{-1} min^{-1}. The high-affinity heparin therefore induces an 1,800-fold increase in the bimolecular rate constant for the thrombin–antithrombin interaction. Rosenberg *et al.* (Jordan *et al.*, 1979) cite this evidence as proof that heparin acts as an anticoagulant by binding to antithrombin. However, some caution should attend the comparative analysis of the kinetic and binding data of Figure 7.11, as the molar ratios of heparin to antithrombin are of an order of magnitude different in the two types of experiments. For example, at heparin concentration of 1 $\times 10^{6}$ M where maximal initial velocity of thrombin inhibition occurs, the heparin:antithrombin molar ratio is 100 for the kinetic data and only 10 for the binding data.

Of particular interest in relation to the results in Figure 7.11 is that at concentrations of high-affinity heparin above 5×10^{-6} M, a decrease in the initial velocity of this inhibitory process occurs. This phenomenon appears to be independent of the nature of the polysaccharide, as addition of either relatively inactive heparin or

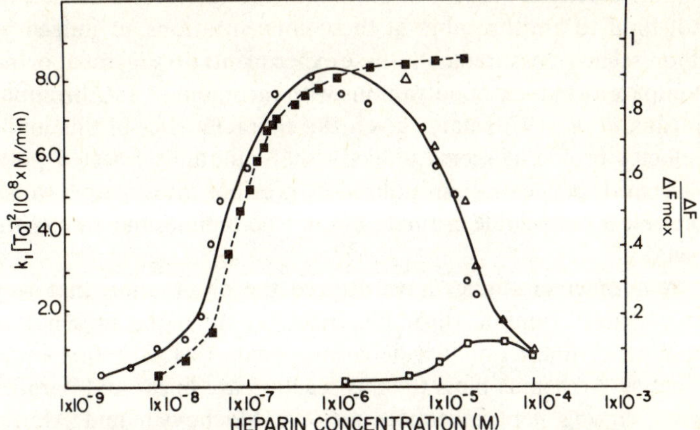

FIGURE 7.11 The extent of binding of heparin species to antithrombin and their ability to accelerate the thrombin–antithrombin interaction. Fluorescence binding curves were obtained with the same fraction of highly active (■——■) heparin utilized in the kinetic analysis. For binding experiments, the concentration of antithrombin employed was 1×10^{-7} M in 0.2M NaCl, 0.01M Tris-HCl, pH 7.5. The binding data are plotted as the fluorescence change at 330 nm for a given concentration of polysaccharide normalized to the projected maximal increase in this parameter (ΔF/ΔFmax). The kinetics of the thrombin–antithrombin interaction were examined in the presence of varying concentrations of highly active (high-affinity) heparin (○——○) or relatively inactive (low-affinity) heparin (□——□). The initial velocity of this process is plotted as $k_1[T_0]^2$ with k_1 evaluated according to the equation $[T]^{-1} - [T_0]^{-1} = -k_1 t$ where $[T_0]$ represents the nominal starting concentration of thrombin and [T] represents levels of residual thrombin observed at varying times t. When highly active heparin is employed, the concentration of enzyme and inhibitor were set at 1×10^{-8} M. With relatively inactive polysaccharide, the level of thrombin and antithrombin utilized were 5×10^{-8} M but the data were normalized to a reactant concentration of 1×10^{-8} M for comparison with the highly active heparin. Also shown is the effect of adding various concentrations of relatively inactive heparin (△——△), to reaction mixtures which contain a level of highly active heparin sufficient to promote an optimal rate of thrombin inactivation. The solvent environment employed in the kinetic analyses was identical with that utilized for obtaining binding curves cited above. (Adapted from Jordan *et al.*, 1979; reproduced with permission of the publishers.)

dextran sulphate to reaction mixtures already containing optimal levels of high-affinity heparin, produced a similar decrease in the initial velocity of enzyme inhibition. Although these materials do not bind to antithrombin at these concentrations, as judged by fluorescence measurements, these experiments do suggest abortive complex formation of polyanions with thrombin or antithrombin. Jordan *et al.* (1979) state, 'given the characteristics of this initial velocity profile, it seems unlikely that randomly selected polysaccharide concentrations utilized by previous investigators would provide a reasonable estimate of the true bimolecular rate of the reaction'.

A number of studies have utilized the observation that with excess antithrombin, thrombin inactivation in the presence or absence of unfractionated heparin appears to follow first order kinetics. Arrhenius plots for this reaction are linear and parallel, with varying heparin concentrations (Machovich and Arányi, 1978). However, the thermodynamics of the reaction are subject to conflicting reports. Studies do suggest that with excess antithrombin the reaction proceeds at a slower rate and may be subject to substrate inhibition by antithrombin. Indeed, Kowalski and Finlay (1979) have recently published some limited data demonstrating that thrombin inactivation by antithrombin in the presence of high-affinity heparin may undergo substrate (antithrombin) inhibition when the antithrombin/thrombin molar ratio was greater than 1.0.

For experiments with antithrombin/thrombin molar ratios in the range of 2.5–6.5, the pseudo first order kinetics of thrombin inactivation in the presence of unfractionated heparin has been theoretically interpreted in terms of heparin activation of thrombin which makes it more susceptible to antithrombin inhibition (Stürzebecher, 1977; Stürzebecher and Markwardt, 1977). This mechanism more suitably described the kinetic data as compared to heparin activation of antithrombin with subsequent inactivation of thrombin or the formation of a trimolecular complex between heparin–thrombin–antithrombin as the primary event. However, a number of observations remain unexplained in terms of invoking a heparin–thrombin interaction as the primary event in thrombin inactivation. A major problem in the use of unfractionated heparin is the variation within a heparin preparation of its affinity with antithrombin and, to some extent, thrombin. There is also the likelihood of substrate inhibition by excess antithrombin. When discussed in these terms, the formation of abortive complexes between antithrombin–thrombin, heparin–antithrombin and heparin–thrombin may occur and therefore decrease the optimal reaction rate of the heparin accelerated inactivation of thrombin by antithrombin. It is likely that these interactions require consideration to explain the apparent anomalies associated with the effects of sequential addition of the reactants in both solution studies (Li *et al.*, 1976), and studies using heparin-linked agarose (Hatton and Regoeczi, 1977) and heparin adsorbed Sepharose-lysine (Hatton *et al.*, 1977).

A speculative mechanism of antithrombin III–heparin–thrombin interaction In the reaction scheme outlined in Figure 7.12 a speculative mechanism describing the range of binary and ternary interactions manifesting sub-optimal and optimal reactions of the antithrombin–heparin–thrombin interaction, which may occur for commercial heparin preparations has been presented. It is clear that sub-optimal interactions do occur, as Jordan *et al.* (1979) have noted that the maximal rate of thrombin inactivation in the presence of a purified high-affinity heparin preparation is 5- to 20-fold higher than the rate observed for normal, commercial, unfractionated heparin preparations.

The central reaction describes a hypothetical distribution of inactive and active heparin. The terms active and inactive used here are merely arbitrary in terms that they may define the most active heparin preparation, all other heparins being relatively inactive compared to the most 'active' pure fraction. Such fractions having homogeneous activity with respect to antithrombin binding have been isolated by Rosenberg *et al.* (Jordan *et al.*, 1979; Rosenberg *et al.*, 1979). The question of whether an 'inactive' heparin exists, which could modify the interaction of active heparin in the thrombin–antithrombin interaction through the formation of abortive complexes as outlined in Figure 7.12, is a conjectural point in the literature. On one hand, Rosenberg *et al.*

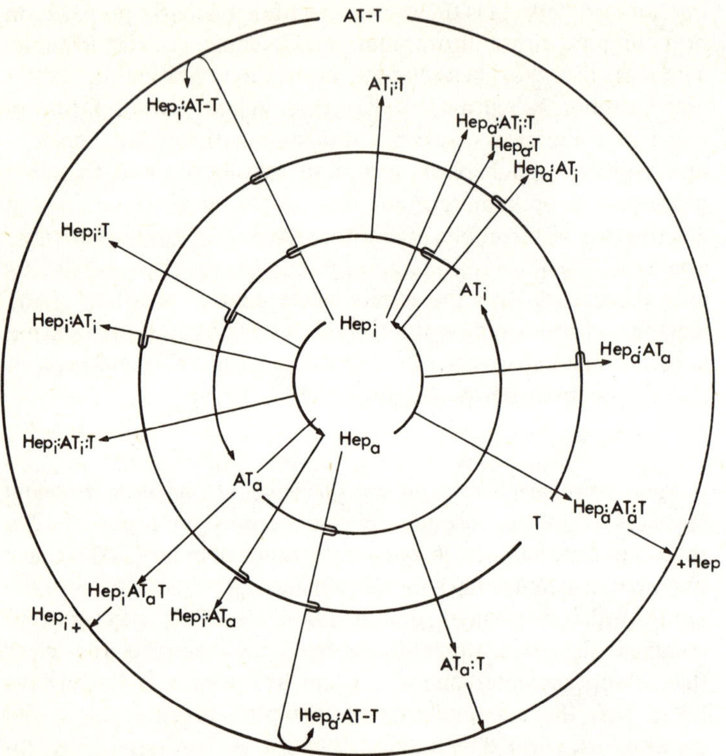

FIGURE 7.12 A speculative mechanism for the range of interactions that may occur for the three component system consisting of thrombin (T), antithrombin III (AT) and heparin (hep). The subscripts i and a denote inactive and active species respectively (see text for details).

(Jordan *et al.*, 1979) suggest that inactive heparin (67% of a low-molecular weight pool) does not bind to antithrombin, as judged by fluorescence measurements and other data, and exhibits low anticoagulant activity (probably due to contamination with highly active heparin species). In contrast, there is a large amount of data which demonstrate that heparin species with low but finite affinity to antithrombin, under physiological conditions, do exist (see for example Höök *et al.*, 1976; Björk *et al.*, 1979). An explanation for these apparent differences may be in the molecular weight of the preparation used in these experiments. Björk *et al.* (1979) and Laurent *et al.* (1978) have characterized low-affinity material from monodisperse fractions of molecular weights 15,000 and 7,800 respectively. It is noteworthy that the proportion of heparin material that does not bind to heparin, under physiological conditions, increases with decreasing molecular weight of heparin preparation. Laurent *et al.* (1978) have found that material with no-affinity to antithrombin constituted 46% of a heparin preparation of molecular weight 5,800. For heparin preparations with molecular weights >13,000, no-affinity heparin species were detectable. It will be taken that normal, unfractionated heparin preparations contain 40–60% low affinity, relatively inactive species (Laurent *et al.*, 1978).

The equilibrium between active heparin and relatively inactive heparin may then conceivably be operated by a number of factors such as 1) decrease in heparin activity through biosynthetic or degradative activity controlling the distribution of antithrombin binding fragments in the heparin chain and 2) a decrease in heparin activity through the formation of complexes with proteins other than thrombin or antithrombin. Whether the inactive and active forms of heparin as depicted in Figure 7.12 are in equilibrium *in vivo* is not known. The fact that active heparin may be inactivated through abortive binary complexes with thrombin or antithrombin, depending on the concentrations of the reactants, has been discussed in preceding sections.

The second circle from the centre represents a hypothetical equilibrium between active and inactive forms of antithrombin. In this case it is conceivable that this reaction may be generated by factors including 1) thrombin proteolysis of antithrombin (Fish and Björk, 1979) and 2) aggregation of antithrombin (Pepper *et al.*, 1977). Furthermore, active antithrombin may be inactivated to some extent by the binding of inactive (low-affinity) heparin to form abortive complexes.

The third phase of the reaction is arbitrarily represented as that with α-thrombin (although not shown this portion also exists in active (α-thrombin) and inactive forms (β-thrombin)). Thrombin may form abortive binary complexes with active or inactive heparin under certain conditions (see earlier sections). The formation of an abortive complex between antithrombin and thrombin has not been demonstrated, although in the sense of the optimal reaction rate in the presence of active heparin, a normal reaction between thrombin and antithrombin, not under the influence of heparin, will be relatively abortive.

There is no general agreement to ascertain which, if any, is the obligatory first reactant with heparin, i.e. thrombin or antithrombin. Rosenberg *et al.* (1980) conclude that it is the direct effect of heparin upon the antithrombin molecule that is responsible for at least 99% of the heparin-dependent acceleration of the thrombin–antithrombin reaction. These conclusions were based on kinetic and binding data of low molecular weight (6,500), highly active heparin discussed earlier. They suggest that interaction of thrombin and heparin, which occurs when the polysaccharide is bound to antithrombin, is kinetically insignificant. Similar conclusions could be invoked to explain the anticoagulant activity registered for the heparin antithrombin binding fragment obtained by heparinase digestion of high-affinity heparin–antithrombin complex. This fragment had a thrombin inactivating capability of 42% as compared to that of a high-affinity heparin ($M = 16{,}500$) preparation (Laurent *et al.*, 1978).

For heparin preparations of molecular weight >6.000, the statistical analysis of the molecular weight dependency of heparin action on the thrombin–antithrombin interaction would indicate that an optimal ternary complex forms, where heparin acts as a surface for activation of antithrombin to neutralize an adjacent thrombin bound to the same heparin molecule (Laurent *et al.*, 1978).

The fourth and final phase of the reaction depicted in Figure 7.12 is the formation of the antithrombin–thrombin covalently linked complex, in which the catalytically active heparin is released from the complex for further participation in other thrombin–antithrombin interactions. Abortive complexes do not reach this stage. Further, the covalent binary complex of antithrombin–thrombin may decrease heparin activity due to association.

7.2.1.5 ANTITHROMBIN III–FACTOR Xa INTERACTION

The potentiation of the inhibitory activity of antithrombin by heparin on Factor Xa is currently regarded as an important site for manifestation of heparin anticoagulant activity. The mechanism of heparin action in this reaction is not clear. Several features of this reaction appear to be distinctly different from heparin action on the thrombin–antithrombin interaction, and these will be discussed briefly.

The molecular weight dependence of heparin activity in the anti-Factor Xa assay appears to be different, particularly for low molecular weight heparins, as compared to the molecular weight dependence in thrombin neutralization, when the activity is expressed on a heparin weight basis, i.e. material of low molecular weight shows considerably higher activity in the anti-Xa assay than by global clotting methods and this activity decreases with increasing molecular weight (Andersson *et al.*, 1976; Barrowcliffe *et al.*, 1977; MacGregor *et al.*, 1979). However, when factor Xa-inactivating potencies of the high-affinity (for AT) heparin are expressed on a heparin molar basis, heparin activity increases continuously with increasing molecular weight of the polysaccharide in a fashion similar to thrombin inactivation by antithrombin and heparin (Thunberg *et al.*, 1979). On this basis it was suggested that factor Xa inactivation occurs by a similar mechanism as thrombin inactivation, where maximal inactivation of the target serine proteinase may require binding of both proteinase and antithrombin molecules to the same heparin chain. Inactivation of factor Xa may, however, be less dependent on the binding of the enzyme to heparin than the inactivation of thrombin.

While the mechanism of anti-Factor Xa neutralization may be similar to thrombin neutralization, several features of the former reaction remain unexplained. Apart from the behaviour of low molecular weight heparin on the reaction, substrate inhibition with excess Factor Xa does not apparently occur (cf excess thrombin Section 7.2.1.4.5) (Marciniak, 1975; see also Rosenberg *et al.*, 1980). A further difference in the factor Xa reaction as compared to thrombin, is that the former requires a phospholipid surface for maximal activity and protection against inhibitory action of antithrombin. Studies have shown that an important feature of the heparin effect in this system is its ability to compete for factor Xa interaction with phospholipid vesicles and the heparin ultimately yields the displacement of Factor Xa from the lipid surface. Heparin can reverse the lipid protection of Factor Xa from antithrombin (Yin, 1974; Walker and Esmon, 1979a) and inhibit prothrombin activation by Factor Xa, calcium and phospholipid (Owen *et al.*, 1974; Walker and Esmon, 1979b). Commercial heparin preparations have been fractionated into high- and low-affinity fractions on Factor Xa-linked Sepharose at low ionic strengths (Walker and Esmon, 1979a). It remains to be determined whether these specific interactions occur at physiological ionic strength.

It is clear that by accelerating the inhibition of Factor Xa in the coagulation cascade, heparin is capable of preventing the development of much larger amounts of thrombin (Wessler and Yin, 1974; Gitel *et al.*, 1977). However, it is questionable whether heparin activity is primarily focused at this stage of the coagulation cascade or the direct neutralization of thrombin. There have been conflicting reports on the activity of heparin against Factor Xa as compared to thrombin and no definite conclusions can be made (Yin and Wessler, 1971; Yin, 1974; Ødegård and Lie, 1978).

7.3 OTHER HAEMOSTATIC COMPONENTS

It is evident that heparin may potentiate the inhibitory action of antithrombin on other serine proteases involved in the haemostatic mechanism outside the coagulation cascade.

The central event of the fibrinolytic system is the conversion of a zymogen (plasminogen) to a serine protease (plasmin) (Figure 7.13). A variety of substances are capable of initiating this transition. These include Factor XIIa via plasminogen proactivator as well as other plasminogen activators found in various tissues, blood and urine. Highsmith and Rosenberg (1974) have provided evidence for the antiplasmin activity of antithrombin and that heparin may dramatically potentiate this inhibition (see also Stürzebecher and Markwardt, 1977). Studies utilizing SDS-electrophoresis have demonstrated that antithrombin forms a 1:1 stoichiometric complex with plasmin that is stable to denaturing agents (Rosenberg, 1977). With respect to the physiological

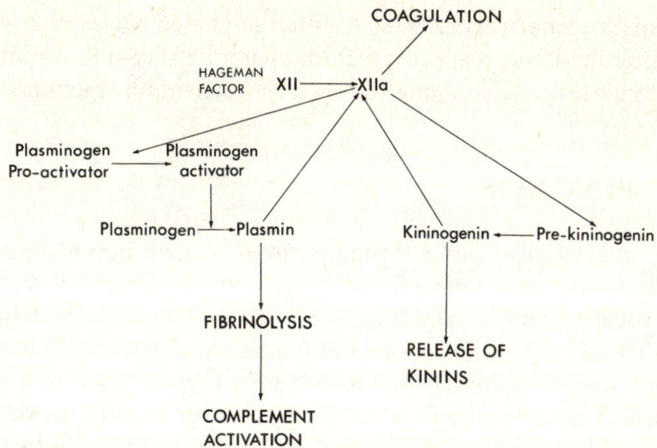

FIGURE 7.13 Interactions between activation of kinin-releasing enzymes (kininogenase or plasma kallikrein), plasmin and the coagulation factors. (From Rocha e Silva, 1978; reproduced with permission of the publishers.)

significance of antiplasmin activity of antithrombin, as compared to a widely assumed major inhibitor of plasmin, α_2-macroglobulin, it has been shown that the complexing of plasmin to α_2-macroglobulin dominates the inhibitor type interactions (Rosenberg, 1977; Semeraro *et al.*, 1978). This situation is reversed in the presence of heparin in that *in vitro* studies suggest that antiplasmin activity is then dominated by antithrombin (Rosenberg, 1977). Current data would suggest that α_2-plasmin inhibitor is the principle antagonist of plasmin action (Moroi and Aoki, 1976; Highsmith *et al.*, 1977). However, its activity as compared to the heparin–antithrombin complex, has not been reported.

Antithrombin has also been shown to have anti-urokinase activity (a plasminogen pro-activator serine protease in urine) and this reaction is greatly enhanced by heparin (Clemmensen, 1978).

With respect to the kinin-generating system, antithrombin slowly neutralizes the action of serine protease kallikrein (Lahiri *et al.*, 1974; Venneröd and Laake, 1975), which converts kininogenin to kinins. Heparin only exerts a mild acceleration of this reaction. Heparin alone is known to activate pre-kininogenin to kininogenin (dos Santos *et al.*, 1978). As other sulphated polysaccharides register this effect, it appears that the charge density of the heparin molecules is a determining factor in this activation reaction.

7.4 PLATELETS

The platelet, also called thrombocyte, is a small non-nucleated body formed when cytoplasmic fragments of cells called megakaryocytes (from bone marrow) are pinched off into the circulation as they age. Under normal circumstances, platelets circulate in the blood stream as single cells, non-adherent to each other or to the vascular endothelial cells which form the inner layer of the vessel walls. When the vessel wall is ruptured, either by injury or by the formation of an atherosclerotic lesion, two main responses are set into action which lead to clot formation. One is a series of cellular alterations of blood platelets, enabling them to aggregate and produce a primary haemostatic plug. The other response follows a series of complex transformations of proteolytic zymogens into active coagulation factors leading to the formation of a fibrin network. The platelet fibrin mass is presumably further consolidated by a mechanism involving the contractile properties of platelets completing the haemostatic process leading to a permanent arrest of bleeding. When the endothelium is ruptured, platelets can adhere to a wide variety of compounds present in the vessel wall or perivascular tissue, such as collagen, basement membrane and microfibril–elastin complex. Of these components, collagen is presumably the most effective, especially as platelet–collagen adhesion initiates an exocytotic secretion of the contents of subcellular storage granules (often referred to as the release reaction) which leads to further platelet aggregation and is primarily responsible for thrombus formation. Platelets which adhere to collagen become subsequently degranulated, thereby releasing a number of substances including ADP, serotonin and a variety of proteins. The release reaction, which can also be induced by, for example, thrombin, is essentially a secretory event accomplished by fusion of storage granules with the plasma membrane, followed by discharge of the granular contents to the cell exterior while the granular membrane remains associated with the plasma membrane. The exact mechanisms leading to exocytosis are not clear. The release of endogenous ADP is held responsible primarily for the subsequent aggregation of platelets leading to haemostatic plug formation. Although the above considerations are fragmentary and incomplete, they serve to illustrate that the sequence of adhesion, release reaction and ADP-induced aggregation is mediated by a number of membrane related events. Platelet membrane lipids (or lipoproteins) also participate in the coagulation process in a non-metabolic way by offering a catalytic surface on which clotting factors interact, resulting in an accelerated formation of thrombin.

The effect of heparin on platelet aggregation is not clear. This has been subject to many conflicting reports, which is probably related to the conditions of study (Zucker, 1977). Recent studies by Salzman *et al.* (1980) have shown that porcine mucosal heparin at a concentration of 10 μg/ml or less induces platelet aggregation in citrated platelet-rich plasma and enhances platelet aggregation and serotonin secretion induced by other agents. This action of heparin was blocked by substances that elevate platelet cyclic AMP and by EDTA but not by inhibitors of arachidonic acid cyclo-oxygenase or by hirudin. The platelet response to heparin does not seem to require the involvement of ADP because it is not inhibited by N-amylthio-5′-AMP or apyrase. Further, Salzman *et al.* (1980) have demonstrated that induction of platelet aggregation and enhancement of aggregation induced by other agents increases with increasing molecular weight of heparin. In high molecular weight fractions, antithrombin affinity appeared to have no relation to platelet activity. In low molecular weight fractions, there was a roughly reciprocal relationship between platelet aggregation activity and antithrombin binding affinity; low-affinity fractions being significantly more reactive towards platelets than high-affinity fractions. However, when platelet-rich plasma is depleted of antithrombin, the low molecular weight heparin fractions of high- or low-antithrombin affinity were equivalent in enhancement of ADP-induced aggregation. These and other results suggested that antithrombin inhibits heparin–platelet interactions. The molecular weight dependent heparin effect may then be explained by the existence of two types of binding site on the heparin molecule, one that binds either to antithrombin or to platelets but has a higher affinity for the former, and a second that binds preferentially to platelets.

A further observation of Salzman *et al.* (1980) is that platelet aggregation by heparin required a plasma cofactor, as heparin fails to aggregate platelets suspended in a buffer after albumin washing and gel filtration. This plasma cofactor is different from the cofactor required for ristocetin.

It is of interest that exogenous heparin, in certain instances, may aggregate platelets and induce thrombocytopenia in some patients. This paradoxical procoagulant activity of heparin is thought to be due to the development of a heparin-dependent IgG platelet aggregating antibody which is directed against an antigen that is contained in heparin preparations (Trowbridge *et al.*, 1978) or an antibody directed against a heparin-platelet component complex (Green *et al.*, 1976). The thrombocytopenic syndromes described in patients receiving heparin may also be explained by the mechanism of Salzman *et al.* (1980) as described above.

The most well established feature of the relationship of platelets to heparin is that platelets have antiheparin activity. This heparin-neutralizing activity has been attributed primarily to (a) low molecular weight protein(s), called platelet factor 4, which is discharged from platelets during the release mechanism associated with platelet aggregation (Niewiarowski *et al.*, 1968). The platelet factor 4 protein is a neutral protein with an isoelectric point of pH 7.6 and a single chain subunit of molecular weight near 9,600 (as determined by sodium dodecyl sulphate electrophoresis technique) (Handin and Cohen, 1976). Until recently, it has been the only known endogenous protein which inhibits heparin anticoagulant activity. (Andersen and Godal (1977) have recently isolated a plasma α_1-acid glycoprotein which also has anti-heparin activity.) The platelet factor 4 probably exists within specific granules of the platelet in a complex with a high molecular weight chondroitin sulphate proteoglycan (Barber *et al.*, 1972). This complex is released from human blood platelets by thrombin. The function of platelet factor 4, apart from its potential in inhibiting endogenous heparin activity, has not been identified. A recent report by Lonky *et al.* (1978) has demonstrated the potent stimulatory action of this protein on human granulocyte elastase; heparin neutralized this stimulatory activity.

The binding specificity of platelet factor 4 to heparin has been studied in several systems. The binding order of connective tissue polysaccharides follows heparin > heparan sulphate > dermatan sulphate > chondroitin 6-sulphate ≃ chondroitin 4-sulphate (Barber *et al.*, 1972; Handin and Cohen, 1976) which suggests that the binding is promoted by high charge densities and is positively correlated with the presence of iduronate. Niewiarowski *et al.* (1979) have also demonstrated that both low-affinity (for AT) heparin and high-affinity heparin show similar affinities for platelet factor 4, which suggests that different sites on the heparin molecule are involved in the binding of these proteins. The binding of platelet factor 4 to heparan suiuue and dermatan sulphate may localize platelet factor 4 at sites on cell surfaces of the vascular wall.

The amino acid composition of human platelet factor 4 indicated 18% basic and 22% acidic amino acid residues assuming all the aspartic and glutamic residues were in the acidic form. The platelet factor 4 thus differs from a 'cationic protein', such as protamine, which also neutralizes heparin (Handin and Cohen, 1976). Modification of lysine residues by guanidination decreased heparin-neutralizing and heparin-binding activity, while modification of arginine residues had no effect (Handin and Cohen, 1976). The importance of lysine residues for heparin binding concurs with the peculiar amino acid sequence at the COOH-terminal region (Deuel *et al.*, 1977) in which pairs of lysine residues are separated by pairs of leucine or isoleucine residues, i.e. –lys–lys–Ile–Ile–lys–lys–leu–leu. Furthermore, a COOH-sequence terminal peptide containing this sequence was able to reduce the heparin-induced inhibition of blood coagulation.

The fact that heparin may also bind to or be taken up by platelets (Shanberge *et al.*, 1976; Horner, 1974; Lindon *et al.*, 1978) has been attributed, in part, to the involvement of platelet factor 4 binding mediated through fibrinogen (Lindon *et al.*, 1978) possibly at the surface membranes or by a pinocytotic mechanism; this does not occur if heparin is complexed with antithrombin (Shanberg *et al.*, 1976). Some of the heparin taken up by platelets is released

(20%) on platelet aggregation and has anticoagulant activity, which suggests that a portion of the heparin binding to platelets is reversible and is not wholly bound to platelet factor 4 (Shanberge *et al.*, 1976).

7.5 LIPOPROTEINS

Lipids exist as soluble complexes with protein (termed lipoproteins) when transported in plasma. Free fatty acids are transported bound to albumin. The phospholipids and neutral lipids are the lipid constituents of the plasma lipoproteins.

The general concept of plasma lipoprotein structure is that they approximate to spherical particles having a central core of triglycerides and/or cholesteryl esters stabilized by an outer monolayer composed of phospholipids, cholesterol and specific apoproteins (apolipoproteins) (for example see Scanu, 1977). The exact nature of the lipoproteins is not known.

There are five major classes of lipoprotein, which are characterized by different densities and vaeying apolipoprotein content (Table 7.4). Each class consists of a spectrum of particles varying in composition as well as size. Changes in density and size of these lipoproteins depend on the overall ratio of lipid core to protein surface. The description of serum lipoproteins as differential ultracentrifugal classes varying in size and density should not be taken to indicate that these are static complexes of rigorously defined stoichiometry; rather, they should be considered as dynamic structures often representing a continuum of subspecies which are the result of intravascular processes of degradation and interconversion (Scanu, 1977).

There is marked lipid heterogeneity expressed in the various classes of lipoprotein as evidenced by the varying composition of alkyl esters of phospholipids, cholesteryl esters and triglycerides. Polar lipids occur in higher quantities in high density lipoproteins and non polar lipids occur to a greater extent in lower density lipoproteins.

The protein composition of the lipoproteins has not yet been completely elucidated, although a large body of experimental

TABLE 7.4
Some Characteristics of Plasma Lipoproteins
(adapted from Scanu (1977) with permission of the publishers)

Class	Abbreviation	Density (gcm^{-3})	Molecular Weight $\times 10^{-6}$	Size (nm)	Apolipoprotein (Apo)	Protein content (weight percent per particle)
Chylomicrons	Chylos	<0.95	10^3–10^4	75–1000	Apo B, Apo A-I, Apo A-II, Apo C	2.0
Very low density	VLDL	0.95–1.006	5–27	30–80	Apo C-I, Apo C-II, Apo C-III Minor components (Apo B, Arginine rich protein (ARP))	8.0
Low density Lipoprotein	LDL	1.006–1.063			Apo B Minor components	—
Subclass	LDL_1	1.006–1.019	2.7–3.5	20.2–30.0	(Apo C-I, Apo C-II, Apo C-III)	—
	LDL_2	1.019–1.063	2.2–2.7	20.0–20.2		21
High density Lipoprotein	HDL	1.063–1.21			Apo A-I, Apo A-II	—
Subclass	HDL_2	1.063–1.125	0.175–0.26	8.5–10.0	Minor components	41
	HDL_3	1.125–1.21	0.15–0.175	7.0–8.5	(Apo C-I, Apo C-II, Apo C-III, Apo B, ARP)	55

evidence favours the concept that apolipoproteins play an essential role as determinants of lipoprotein structure. There are essentially three major apolipoproteins, namely A, B and C. Apolipoprotein A, which is considered to be of two types, Apo A-I and Apo A-II, is the major constituent of the HDL and is also found in chylomicrons. Apolipoprotein B is less well characterized and is the principal constituent of the LDL and is also found in chylomicrons, VLDL and HDL. Apolipoprotein C has been separated into three distinct fractions, namely Apo C-I, Apo C-II and Apo C-III. These are most abundant in VLDL, but are also found in chylomicrons, LDL and HDL.

There have been numerous studies that have shown that plasma lipoproteins may interact with glycosaminoglycans (particularly heparin) *in vitro*, although the functional significance of these interactions is far from clear. These interactions have been studied by a number of techniques, including the early studies by Bernfeld utilizing nephelometry and free-boundary electrophoresis (see review by Bernfeld, 1966), by equilibrium-binding of lipoproteins to polysaccharide-substituted agarose gel (Iverius, 1972; Srinivasan *et al.*, 1975), by precipitation techniques (Srinivasan *et al.*, 1975; Nakashima *et al.*, 1975), by fluorescence spectroscopy with pyrene-labelled lipoproteins (Nakashima *et al.*, 1975) and the binding of [^{3}H]-heparin to plasma lipoproteins using gel chromatography to fractionate the heparin–lipoproteins complexes (Pan *et al.*, 1978).

Variable results between laboratories have been obtained on the binding of heparin to various lipoproteins. It is likely that the use of preparations of lipoproteins from different sources and methods of preparation of lipoprotein may account, in part, for these variations.

Minor apolipoprotein components may be responsible for binding and may exhibit some variability with various preparations. In this regard, the studies by Shelburne and Quarfordt (1977) have shown that heparin has a particularly high reactivity towards arginine-rich apolipoproteins obtained from human VLDL, although it is only a minor component of this lipoprotein. It appears that this component is primarily responsible for the binding of heparin to this lipoprotein. Those VLDL fractions with no arginine-rich-protein do not bind to heparin, as measured by binding to heparin–agarose gel at physiological ionic strength and in the absence of divalent cations. The isolated arginine-rich apolipoprotein and whole lipoproteins which were treated with phenylglyoxal with a resulting alteration of the major portion of the arginine residues were not able to bind to a heparin-affinity column.

Iverius (1972) first established the use of glycosaminoglycan bound-Sepharose for interaction studies. He found that sulphated glycosaminoglycans bound to the gel matrix possessed a strong capacity to bind VLDL and LDL. The binding was highly dependent on ionic strength in that at very low and very high ionic strengths no binding was observed and only occurred in the middle region. The release of half the amounts of LDL or VLDL from gels of heparin, dermatan sulphate, heparan sulphate, and chondroitin 4-sulphate required ionic strengths of 0.26, 0.15, 0.09 and 0.08 respectively. HDL did not bind. The binding could be abolished by N-acetylation of the VLDL or LDL, suggesting an importance for the lysine residues. Divalent cations were not required for the interaction. In view of the similar binding of the LDL and VLDL to heparin it was suggested that the apolipoprotein B was, in part, responsible for binding to heparin.

Recent studies by Pan *et al.* (1978) have identified the formation of a soluble complex between [^{3}H]-heparin and LDL by gel chromatography. However, in these studies no binding of HDL or VLDL occurred. The order of competition of [^{3}H]-heparin for lipoprotein in a complex formed by other glycosaminoglycans was heparin > heparan sulphate > chondroitin sulphate > keratan sulphate > dermatan sulphate. There was also preference for higher molecular weight species of heparin as compared to lower molecular weight species. However, the binding of [^{3}H]-heparin to LDL in this system was relatively low (25%) when measured under physiological conditions. The LDL was treated with ethoxyformic anhydride for histidine modification, or acetic anhydride and succinic anhydride for lysines and cyclohexanedione for arginine residues; in each case there was a significant loss in heparin binding capacity, suggesting that the various basic amino acids are involved in the binding and/or that basic amino acids are necessary to maintain the proper conformation of LDL. In view of the variability of heparin interaction, it would appear that no conclusions can yet be drawn as to the exact mechanism of the interaction. However, on the basis of observations gained from modification of basic amino acid residues, it would appear likely that the binding of heparin to lipoproteins involves binding between basic amino acid groups in the apolipoproteins and acidic groups in the polysaccharide to form soluble complexes. It is, of course, possible that alteration of some of the basic amino acids changes the conformation of the interacting lipoproteins in such a way as to prevent their binding to heparin.

Although divalent ions do not appear to be required for the formation of soluble heparin–lipoprotein complexes (Iverius, 1972; Pan *et al.*, 1978), they may stabilize such interactions of heparin and LDL and VLDL in a very narrow range of metal ion concentration (0.02–0.04M) in the absence of other salts (Srinivasan *et al.*, 1975); Mn^{2+} was more active in this respect than either Mg^{2+} or Ca^{2+}. It is to be noted, however, that formation of divalent ion mediated heparin–lipoprotein complexes under physiological conditions only occurs to any significant extent with Mn^{2+}. Srinivasan *et al.* (1975) found that the NaCl concentration required for 50% dissociation of the complexes was 0.03, 0.09 and 0.19M in the presence of Ca^{2+}, Mg^{2+} and Mn^{2+} respectively. Whereas all three cations promoted complex formation with LDL and VLDL, their specificities in forming complexes with HDL were markedly different in that only Mn^{2+} was able to induce complexes between heparin and HDL (Srinivasan *et al.*, 1975). The amount of complex formation with HDL would appear to be dependent on the quantities of HDL_2 and HDL_3 which have different propensity to form heparin complexes (Srinivasan *et al.*, 1975; Srinivasan *et al.*, 1978). It is to be noted, once again, that complex formation with HDL occurred only at low ionic strengths and its physiological significance is therefore unclear. The differential precipitation of the various lipoproteins with divalent cations (when used at high concentrations) has been used for quantitative estimates of particular lipoprotein classes in blood (for review see Burstein and Scholnick, 1973).

The nature of the interaction between heparin and lipoprotein in the presence of divalent cations appears to be different to those soluble complexes formed in the presence of monovalent cations. In particular, N-acetylated LDL may form complexes with heparin (Srinivasan *et al.*, 1975) and synthetic sulphated polysaccharides (Bernfeld and Kelley, 1963; Nishida and Cogan, 1970) in the presence of divalent cations. In view of this and other evidence it is viewed that added divalent cations bridge the anionic groups of heparin and LDL (possibly through protein and/or phospholipids) thereby producing cross links necessary for forming insoluble complexes. Mookerjea (1978) has recently shown that in the presence of phosphorylcholine, interaction of heparin with serum lipoproteins requires considerably smaller Ca^{2+} concentrations (of the order of 0.01M). This pattern of interaction was not evident for purified lipoproteins, which suggested the requirement of a serum factor for the phosphorylcholine to lower the requirement for Ca^{2+}. Although these reactions were not performed at physiological ionic strength, they do point to the potential importance of phospholipids in heparin–lipoprotein interactions. At the present time, although divalent cations may play a role in heparin–lipoprotein interactions *in vivo*, this has yet to be convincingly demonstrated in model systems performed under physiological conditions.

Recent studies on the effect of glycosaminoglycans on lipoprotein metabolism have pointed towards some functional possibilities of these interactions. LDL, the major cholesterol carrying lipoprotein in human plasma, delivers its cholesterol to human fibroblasts in tissue culture by a receptor-mediated process (Brown *et al.*, 1975a). The LDL binds with high affinity to a specific cell surface receptor, which is then taken up by the cell presumably through endocytosis and becomes incorporated into lysosomes, where its protein and cholestryl ester components are hydrolyzed (for review see Brown and Goldstein, 1979). The nature of the receptors is not exactly known but has been recently equated with protein coated pits on the fibroblast surface (Brown and Goldstein, 1979). The binding of LDL to its cell-surface receptor of normal human fibroblasts can be reversed, as measured by the release of $[^{125}I]$-labelled LDL, by adding heparin to the culture medium. It is probable that heparin acts by competing with the LDL receptor for binding to LDL and raises the possibility that cell surface glycosaminoglycans may in fact form part of the receptor structure. Dermatan sulphate at higher concentrations produced a similar release effect whereas chondroitin sulphate was inactive (Goldstein *et al.*, 1976). The binding of lipoproteins to the cell surface indeed shows analogous behaviour to that observed in model heparin–agarose systems. Mahley and Innerarity (1977) have shown that lipoproteins containing arginine-rich apolipoproteins may compete effectively with [^{125}I]-LDL for binding internalization and degradation by human fibroblasts.

A role for glycosaminoglycans, especially dermatan sulphate, in development of the atherosclerotic lesion has been postulated (Iverius, 1973; Ross and Harker, 1976; Srinivasan *et al.*, 1972). The formation of the athersclerotic plaque, as a result of repeated or chronic injury to the arterial wall (Ross, 1979), is believed to be promoted by the interaction between lipoprotein (particularly LDL, the major atherogenic lipoprotein) and internal glycosaminoglycans. It is unlikely that heparin-like polysaccharides have any significant effect in this regard, as there is no evidence for the presence of heparin in the extracellular space of the arterial wall and that heparan sulphate does not bind lipoproteins at physiological ionic strength as measured in model systems. However, it is of particular clinical interest that exogenous sulphated glycosaminoglycans administered to animals have alleged anti-atherosclerotic effects. Maximal effects were obtained with heparin. This behaviour has been attributed to controlling the interaction of atherogenic lipoproteins with their arterial receptors, possibly glycosaminoglycans (Day *et al.*, 1975; Day, 1976).

7.6 LIPOPROTEIN LIPASE

Lipoprotein lipase appears to be a key enzyme in the mammalian lipid metabolism. It hydrolyzes the triglycerides contained in VLDL and chylomicrons and thereby facilitates uptake of their constituent fatty acids by the extrahepatic tissues of the body. Hahn (1943) first demonstrated that intravenous injection of heparin results in an accelerated rate of removal of triglycerides from plasma. This so called clearing of plasma was shown to be due to the action on plasma triglyceride of lipolytic enzymes which are released from tissues by heparin (Korn, 1955). These enzymes are normally not present in the blood but are involved in the hydrolysis of chylomicrons and VLDL triglycerides occurring on the lumenal surfaces of the capillary endothelial cells of the extrahepatic tissues. As a result of enzyme action, free fatty acids become available to various tissues to be reesterified or oxidized and the lipoproteins, now depleted of most of their triglyceride moiety form higher density lipoproteins, which are probably degraded in the liver.

At least two lipases are released into the blood after heparin injection. Lipoprotein lipases from several extrahepatic tissues (for review see Borensztajn, 1977) appear to have similar properties. This particular enzyme has a requirement for apolipoprotein C-II as a cofactor, is inhibited by high concentrations of NaCl and has alkaline pH optimum. Purified lipoprotein lipase from adipose tissue (Greten and Walter, 1973; Bensadoun *et al.*, 1974) heart (Ehnholm *et al.*, 1975; Twu *et al.*, 1976) and milk (Egelrud and Olivecrona, 1972; Iverius, 1972) share most of these properties. The identification of the endothelial surface as the site of action of lipoprotein lipase suggests that the enzyme might be synthesized in the endothelial cell. However, no data are presently available to support this view. Most of the evidence indicates that the enzyme is synthesized in parenchymal cells from where it is secreted and eventually transported to the endothelial surface (Cryer *et al.*, 1975; Nilsson-Ehle *et al.*, 1976).

The other major lipase released into the blood by heparin probably originates in the liver (Grèten *et al.*, 1972; Assman *et al.*, 1973; Krauss *et al.*, 1973; Brown *et al.*, 1975b). *In vitro* it can be distinguished from the extrahepatic lipoprotein lipase, since its activity is not enhanced by any of the known apolipoproteins and is not inhibited by high concentrations of sodium chloride. This lipase also hydrolyzes phospholipids and partial glycerides (Brown *et al.*, 1975b) and may be identical with a phospholipase that is released from hepatocyte plasma membranes on incubation with heparin (Waite and Sisson, 1973).

An extrahepatic phospholipase may also be released by heparin (same as lipoprotein lipase?) which has hydrolytic activity towards phosphatidyl ethanolamine and phosphatidylcholine (Eisenberg and Schurr, 1976; Stocks and Galton, 1978).

If it is assumed that endogenous heparin from the mast cells, released into the circulation acts in the same way as exogenous heparin it would appear that the heparin needs to be depolymerized to its constituent chains for activity. Horner (1972) has shown that the macromolecular heparin from rat skin exhibits poor lipoprotein lipase-releasing activity *in vivo*. It is only when this heparin is depolymerized, by incubation with the 100,000 × g supernatant medium from a sonicated homogenate of rat small intestine, to yield essentially heparin chains, does it gain clearing activity.

In addition to the lipoprotein lipase-releasing activity from endothelial cell surfaces, exogenous heparin has also been implicated in the process of lipoprotein–lipase interaction with its substrates. The fact that heparin can bind to lipoprotein lipase has been demonstrated by studies which showed that insolubilized heparin–agarose readily binds lipoprotein lipase (Iverius, 1971). This binding has been widely used in the purification of lipoprotein lipase from various sources (Egelrud and Olivecrona, 1972; Hernell *et al.*, 1975; Grèten and Walter, 1973; Bensadoun *et al.*, 1974; Ehnholm *et al.*, 1975; Twu *et al.*, 1976; Bengtsson and Olivecrona, 1977).

The interaction between lipoprotein lipase and heparin has been studied in some detail. In contrast to the heterogeneous heparin-binding characteristics to antithrombin, lipoprotein lipase binds with high affinity to all molecules in heparin preparations; the molecular species that have low affinity for antithrombin and little or no anticoagulant activity, could not be distinguished from heparin which has high affinity for antithrombin, as measured by affinity chromatography on immobilized lipoprotein lipase (Bengtsson *et al.*, 1977). Moreover these workers found that low-affinity heparin efficiently released both lipoprotein lipase and hepatic lipase on intravenous injection into rats. It is of particluar interest both in terms of physiological function and therapeutic value that heparin preparations may contain low-affinity heparin that affects lipoprotein lipase but lacks effect on blood coagulation.

It also seems possible that, in view of the wide concentration of NaCl (0.6–1.0M) required to elute heparin from lipoprotein lipase–Sepharose (Bengtsson and Olivecrona, 1977; Olivecrona and Bengtsson, 1978), there exist subfractions within a heparin preparation which have relatively higher and lower affinity for lipoprotein lipase.

The effect of heparin on lipoprotein lipase activity has been subject to conflicting evidence and no definite conclusions can be drawn at present (Iverius *et al.*, 1972; Olivecrona *et al.*, 1977; Borensztajn, 1977). Iverius *et al.* (1972) found that the effect of heparin on the activity of bovine milk lipoprotein lipase varies with the mode of substrate activation, the ionic strength of the assay medium and the purity of the enzyme. Interestingly, they found that the activity of the crude enzyme was extensively inhibited in the presence of serum; the inhibitory effect of serum was completely abolished by the addition of small amounts of heparin. It would appear then that heparin acts merely to stabilize the enzyme in blood. The polysaccharide does not actually stimulate the enzyme, but rather stimulates a system which is otherwise under suboptimal conditions. It has been suggested that heparin or endogenous cell surface heparan sulphate may also play a role in linking the enzyme to the tissue cell surface and therefore facilitate its action on lipid substrates in that region (Olivecrona *et al.*, 1977). It is also conceivable that the heparin-like polysaccharides may also link lipoprotein substrates to the cell surface and participate in facilitating the reaction of enzyme and substrates.

7.7 COMPLEMENT SYSTEM

The term 'complement' is applied to a system of enzymes, looked upon as one of the enzyme–cascade systems occurring in normal serum that are activated, characteristically, by antigen–antibody interaction. The complement system plays a role in host immunity to various microbes and involves the binding of complement to antibodies attached to surface antigens of bacteria and viruses; the microorganisms are subsequently lysed or those with antibody-coated surfaces are more readily ingested by host phagocytes.

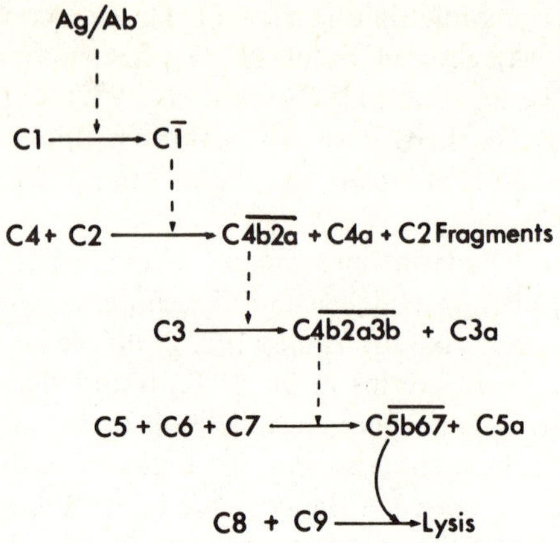

Ag/Ab

C1 ⟶ $C\overline{1}$

C4 + C2 ⟶ $C\overline{4b2a}$ + C4a + C2 Fragments

C3 ⟶ $C\overline{4b2a3b}$ + C3a

C5 + C6 + C7 ⟶ $C\overline{5b67}$ + C5a

C8 + C9 ⟶ Lysis

FIGURE 7.14 The reaction sequence in immune haemolysis (see text for details).

The complement system consists of eleven plasma glycoproteins (Clq, Clr, Cls, C2, C3, C4, C5, C6, C7, C8, C9). The system may be activated by complexes of antigen and antibody (IgM, IgG1, IgG2 or IgG3) or by aggregated human immunoglobulins. The 'classic' complement reaction sequence is shown in Figure 7.14. Activation of the complement 'cascade' is initiated by the binding of the Clq subunit of the first component, Cl, to the immunoglobulin F_c region of the antibody–antigen complex. The active site of the esterolytic activity on $C\overline{1}$ is on the Cls subunit (the bar designates enzyme activation). Activation of the Cl complex requires inhibition of the $C\overline{1}$ inhibitor and the presence of calcium. Once the $C\overline{1}$ esterase activity exists, the second stage of the cascade involves $C\overline{1}$ activity on C4 which is cleaved into at least two fragments; a large fragment C4b and a small fragment C4a. The C4b fragment has an active site that can combine with a receptor on the cell surface. While a small portion of C4b is bound, the major portion of the C4b fragments become inactive and remain in the blood serum. Little is known of the fate of the C4a fragment. The next step involves the adsorption of the complement protein C2 to the cell-bound C4b which requires the presence of Mg^{2+}. Following adsorption, the C2 molecule is cleaved by a neighbouring $\overline{C1s}$ enzyme, also requiring Mg^{2+} ions, into two large fragments, one of which, C2a, becomes bound to C4b so that the complex $\overline{\mathrm{C4b2a}}$ is an enzyme. This enzyme functions as a convertase enzyme for complement component C3 to form two fragments. The small fragment, C3a is released into the fluid phase and plays a role as a mediator of inflammation. The large fragment C3b becomes bound to a receptor on the cell surface. The cellular attachment of C3b then has the potential for creating the complex, $\overline{\mathrm{C4b2a3b}}$ that exhibits a new enzyme activity which then cleaves C5. The C3b fragments that become bound to other sites on the cell surface play an important role as promotors of phagocytosis. The small C5a fragment, is released into the fluid phase and also plays a role as a mediator of inflammation. The larger fragment, C5b, may remain on the $\overline{\mathrm{C4b2a3b}}$ enzyme as it combines with the complement proteins C6 and C7 or it may form a complex only with C6 before it dissociates from the enzyme and combines with C7 in the fluid phase. After the $\overline{\mathrm{C5b67}}$ complex is formed it binds to the cell membrane and is capable of reacting with C8 and C9. In this circumstance, cell destruction ultimately occurs. The activation of C3 by way of antibody, C1, C4 and C2 is designated as the classical pathway. Besides $\overline{\mathrm{C4b2a}}$ (the classical C3-convertase), there are a number of other C3 splitting enzymes that can produce complement activation starting at this point in the cascade. These include serine–histidine esterases such as trypsin, plasmin and thrombin and in the case of the latter two, this action may have physiological significance since C3 is always found on fibrin clotted in fresh serum. There is, however, a further physiological mechanism for generating a C3-convertase which is believed to be of importance *in vivo*. An alternative pathway for the activation of the complement pathway through the C3 to C9 sequence is the so called properdin system (Figure 7.15). It is involved in the neutralization of infectivity of some viruses and has bactericidal activity for certain gram negative bacteria. The initiators of this bypass sequence include endotoxins (lipopolysaccharide of cell walls of gram negative bacteria), zymosan (polysaccharide from yeast cell walls) and aggregated antibodies of the Ig classes (e.g. human IgG-4, IgA-1, IgA-2, IgE and guinea pig $\gamma 1$). In this pathway various initiators apparently react with the properdin system and bring

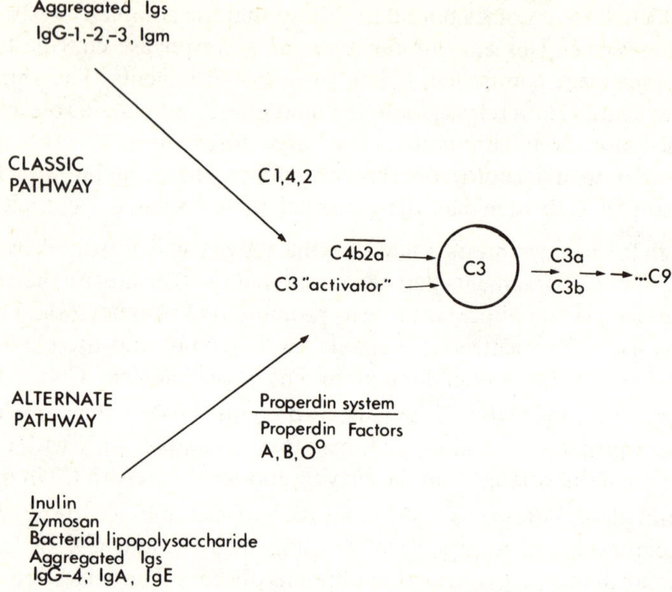

FIGURE 7.15 The properdin system as the alternative pathway for activating C3 and late acting components. Different classes of immunoglobulins activate the two pathways. Cobra venom can also bring about cleavage of C3 by activating factor B of the properdin system. (From Eisen, 1974; reproduced with permission of the publishers.)

about the formation of a proteolytically active substance (C-3 activator) that cleaves C3 in the same way as $\overline{\text{C4b2a}}$. The properdin system is now known to consist of properdin itself and at least three other serum proteins called A, B and the 0° factor. Activation of factor B (also called C3 proactivator) appears to be directly involved in generating the C3-cleaving activity. It remains to be established as to whether an antibody is required to initiate this alternative pathway. Since the properdin system also spares the utilization of C1, C4 and C2, it does not require Ca^{2+}.

Commercial heparin has long been known to possess anti-complementary activity (Ecker and Gross, 1929; Ecker and Pillemer, 1941; Osler *et al.*, 1959). In order to understand the mechanism of this heparin action, several sites of the complement pathway have been studied. Heparin and other polyanions have been shown to directly inhibit $C\bar{1}$ by interaction with the binding site of $C\bar{1}$ to antibody–antigen complexes, namely C1q. The inhibitory effect of the polyanions increased with their increasing charge density. Heparin was shown to have no effect on purified C4 and C2. A further site of inhibition by heparin on $C\bar{1}$ is the inhibition of C4 and C2 consumption by $C\bar{1}$ (Raepple *et al.*, 1976; Loos *et al.*, 1976b; Strunk and Colten, 1976). However, variable results have been obtained where heparin exerted no effect (Ogston *et al.*, 1976) or varying degrees of inhibition of C2:C4 which range from 2–3 (Loos *et al.*, 1976b) to 100 Strunk and Colten, 1976)). The inhibitory effect of heparin in this reaction noncompetitive (Strunk and Colten, 1976) and has been suggested to interfere with the binding of C4 and C2 to $C\overline{1s}$ at a site on the enzyme distinct from the active site (Loos *et al.*, 1976b). At present, it is difficult to draw quantitative conclusions from these experiments in view of the different systems employed and lack of standard quantitative definition of the concentrations of components used. These considerations also apply to other reported actions of heparin that have included the demonstration of inhibition of C2 uptake by membrane-bound C4b, probably through sequestering of Mg^{2+} ions (Loos *et al.*, 1976a), potentiation of the $C\bar{1}$ esterase inhibitor (Rent *et al.*, 1976), the inhibition of formation or binding of the trimolecular complex $C\overline{5b67}$ (Baker *et al.*, 1975) and the sensitive inhibition of C3b-dependent C3 convertase ($C\overline{3bBb}$) by both macromolecular rat peritoneal mast cell heparin, commercial heparin and mast cell granules (Weiler *et al.*, 1978).

A paradoxical effect of the anticomplement activity of heparin has been observed *in vivo* where interactions between heparin and protamine in normal human serum have been found to provoke virtually complete depletion of the total haemolytic complement (C) activity (Rent *et al.*, 1975). This reaction does not appear to occur *in vitro* (Strunk and Colten, 1976). It has been suggested that C-activation by the interactions between certain polyanions and polycations may play a role in the initiation of inflammatory reactions (Claus *et al.*, 1977).

REFERENCES

Abilgaard, U., *Scand. J. Clin. Lab. Invest.* **21**, 89 (1968).

Amiconi, G., Zolla, L., Vecchini, P., Brunori, M. and Antonini, E., *Eur. J. Biochem.* **76**, 339 (1977).

Andersen, P. and Godal, H. C., *Haemostas.* **6**, 339 (1977).

Andersson, L.-O., Barrowcliffe, T. W., Holmer, E., Johnson, E. A. and Sims, G. E., *Thromb. Res.* **9**, 575 (1976).

Andersson, L.-O., Engman, L. and Henningsson, E., *J. Immunol. Methods* 14, 271 (1977).

Assman, G., Krauss, R. M., Fredickson, D. S. and Levy, R. I., *J. Biol. Chem.* **248**, 1992 (1973).

Baker, P. J., Lint., T. F., McLeod, B. C., Behrends, C. L. and Gewurz, H., *J. Immunol.* **114**, 554 (1975).

Barber, A. J., Käser-Glanzmann, R., Jakábová, M. and Lüscher, E. F., *Biochim. Biophys. Acta* **286**, 312 (1972).

Barlow, G. H., Sanderson, N. D. and McNeill, P. D., *Arch. Biochem. Biophys.* **84**, 518 (1961).

Barrowcliffe, T. W., Johnson, E. A., Eggleton, C. A. and Thomas, D. P., *Thromb. Res.* **12**, 27 (1977).

Barrowcliffe, T. W., Johnson, E. A. and Thomas, D. P., *Brit. Med. Bull.* **34**, 143 (1978).

Bartl, K., *Thromb. Res.* **13**, 1141 (1978).

Bengtsson, G. and Olivecrona, T., *Biochem. J.* **167**, 109 (1977).

Bengtsson, G., Olivecrona, T., Höök, M. and Lindahl, U., *FEBS Lett.* **79**, 59 (1977).

Bensadoun, A., Ehnholm, C., Steinberg, D. and Brown, W. V., *J. Biol. Chem.* **249**, 2220 (1974).

Bernfeld, P., in *The Amino Sugars*, E. A. Balazs and R. W. Jeanloz (eds.), Academic Press, New York, Vol. 2B, p252 (1966).

Bernfeld, P. and Kelley, T. F., *J. Biol. Chem.* **238**, 1236 (1963).

Biggs, R., Denson, K. W. E., Akman, N., Borrett, R. and Hadden, M., *Br. J. Haematol.* **19**, 283 (1970).

Björk, I., Danielsson, Å., Fish, W. W., Larsson, K., Lieden, K. and Nordenman, B., in *The Physiological Inhibitors of Coagulation and Fibrinolysis*, D. Collen, B. Wiman and M. Verstraete (eds.), Elsevier/North Holland Biomedical Press, Amsterdam p67 (1979).

Björk, I. and Nordenman, B., *Eur. J. Biochem.* **68**, 507 (1976).

Blaskó, G., Bédi, J., Machovich, R. and Pálos, L. Á., *Haemotologia* **11**, 163 (1977).

Borensztajn, J., in *The Biochemistry of Atherosclerosis*, A. M. Scanu, R. W. Wissler and G. S. Getz (eds.), Marcel Dekker, New York and Basel, p231 (1977).

Borsodi, A. and Machovich, R., *Biochim. Biophys. Acta* **566**, 385 (1979).

Braswell, E., *Biochim. Biophys. Acta* **158**, 103 (1968).

Brinkhous, K. M., Smith, H. P., Warner, E. B. and Seegers, W. H., *Am. J. Physiol.* **125**, 683 (1939).

Brown, M. S., Faust, J. R. and Goldstein, J., *J. Clin. Invest.* **55**, 783 (1975a).

Brown, M. S. and Goldstein, J. L., *Proc. Natl. Acad. Sci. USA* **76**, 3330 (1979).
Brown, W. V., Shaw, W., Baginsky, M., Boberg, J. and Augustin, J., in *Lipoprotein Metabolism*, H. Grèten (ed.), Springer-Verlag, Heidelberg, p2 (1975b).
Burstein, M. and Scholnick, H. R., *Adv. Lipid. Res.* **11**, 67 (1973).
Carlström, A. S., Lieden, K. and Björk, I., *Thromb. Res.* **11**, 785 (1977).
Cifonelli, J. A., *Carbohydr. Res.* **37**, 145 (1974).
Cifonelli, J. A. and King, J., *Carbohydr. Res.* **21**, 173 (1972).
Claus, D. R., Siegel, J., Petras, K., Skor, D., Osmand, A. P. and Gewurz, H., *J. Immunol.* **118**, 83 (1977).
Clemmensen, I., *Thromb. Haemostas.* **39**, 616 (1978).
Collen, D., Schetz, J., de Cock, F., Holmer, E. and Verstraete, M., *Eur. J. Clin. Invest.* **7**, 27 (1977).
Cryer, A., Davies, P. and Robinson, D. S., in *Blood and Arterial Wall in Atherogenesis and Arterial Thrombosis*, J. G. A. J. Hautvast, R. J. J. Hormus, F. van den Haar (Eds.), Leiden, Brill, p102 (1975).
Damus, P. S., Hicks, M. S. and Rosenberg, R. D., *Nature* **246**, 355 (1973).
Damus, P. S. and Rosenberg, R. D., *Methods Enzymol.* **45**, 653 (1976).
Danielsson, Å. and Björk, I., *Eur. J. Biochem.* **90**, 7 (1978).
Danishefsky, I., *Fed. Proc.* **36**, 33 (1977).
Danishefsky, I., Ahrens, M. and Klein, S., *Biochim. Biophys. Acta* **498**, 215 (1977).
Danishefsky, I., Tzeng, F., Ahrens, M. and Klein, S., *Thromb. Res.* **8**, 131 (1976).
Day, C. E., *Atherosclerosis* **25**, 199 (1976).
Day, C. E., Powell, J. R. and Levy, R. S., *Artery* **1**, 126 (1975).
Deuel, T. F., Keim, P. S., Farmar, M. and Heinrikson, R. L., *Proc. Natl. Acad. Sci. USA* **74**, 2256 (1977).
Di Ferrante, N. and Popenoe, E. A., *Carbohydr. Res.* **13**, 306 (1970).
Ecker, E. E. and Gross, P., *J. Infect. Dis.* **44**, 250 (1929).
Ecker, E. E. and Pillemer, L., *J. Immunol.* **40**, 73 (1941).
Egelrud, T. and Olivecrona, T., *J. Biol. Chem.* **247**, 6212 (1972).
Ehnholm, C., Shaw, W., Grèten, H. and Brown, W. V., *J. Biol. Chem.* **250**, 6756 (1975).
Einarsson, R. and Andersson, L.-O., *Biochim. Biophys. Acta* **490**, 104 (1977).
Eisen, H., *Immunology*, Harper and Row, Publishers, Inc., Maryland (1974).
Eisenburg, S. and Schurr, D., *J. Lipid. Res.* **17**, 578 (1976).
Engelberg, H., *Fed. Proc.* **36**, 70 (1977).
Fish, W. W. and Björk, I., *Eur. J. Biochem.* (in press) (1979).
Fish, W. W., Orre, K. and Björk, I., *FEBS Lett.* **98**, 102 (1979).
Foster, A. B., Harrison, R., Inch, T. D., Stacy, M. and Webber, J. M., *J. Chem. Soc.* p2279 (1963).
Fransson, L.-Å., *Carbohydr. Res.* **62**, 235 (1978).
Fransson, L.-Å. and Lewis, W., *FEBS Lett.* **97**, 119 (1979).
Furugren, B., Andersson, L.-O. and Einarsson, R., *Arch. Biochem. Biophys.* **178**, 419 (1977).
Gentry, P. W. and Alexander, B., *Biochem. Biophys. Res. Commun.* **50**, 500 (1973).
Gitel, S. N., in *Heparin: Structure, Function and Clinical Implications*, R. A. Bradshaw and S. Wessler (eds.), Plenum Press, New York and London

(*Adv. Exp. Med. Biol.* Vol. **52**) p243 (1975).

Gitel, S. N., Stephenson, R. C. and Wessler, S., *Proc. Natl. Acad. Sci. USA* **74**, 3028 (1977).

Glimelius, B., Bush, C. and Höök, M., *Thromb. Res.* **12**, 773 (1978).

Goldstein, J. L., Basu, S. K., Brunschede, G. Y. and Brown, M. S., *Cell* **7**, 85 (1976).

Green, D., Harris, K., Reynolds, N., Roberts, M. and Patterson, R., *J. Lab. & Clin. Med.* **91**, 167 (1978).

Grèten, H. and Walter, B., *FEBS Lett.* **35**, 36 (1973).

Grèten, H., Walter, B. and Brown, W. V., *FEBS Lett.* **27**, 306 (1972).

Griffith, M. J., Kingdom, H. S. and Lundblad, R. L., *Biochem. Biophys. Res. Comm.* **83**, 1198 (1978).

Griffith, M. J., Kingdom, H. S. and Lundblad, R. L., *Arch. Biochem. Biophys.* **195**, 378 (1979).

Handin, R. I. and Cohen, H. J., *J. Biol. Chem.* **251**, 4273 (1976).

Hahn, P. F., *Science* **98**, 19 (1943).

Hatton, M. W. C., Kaur, H. and Regoeczi, E., *Biochem. Soc. Trans.* **5**, 1443 (1977).

Hatton, M. W. C. and Regoeczi, E., *Thrombin. Res.* **10**, 645 (1977).

Hernell, O., Egelrud, T. and Olivecrona, T., *Biochim. Biophys. Acta* **381**, 233 (1975).

Hiebert, L. M. and Jaques, L. B., *Thromb. Res.* **8**, 195 (1976a).

Hiebert, L. M. and Jaques, L. B., *Artery* **2**, 26 (1976b).

Highsmith, R. F. and Rosenberg, R. D., *J. Biol. Chem.* **249**, 4335 (1974).

Highsmith, R. F., Wierich, C. J. and Burnett, C. J., *Biochem. Biophys. Res. Commun.* **79**, 648 (1977).

Holmer, E., Söderström, G. and Andersson, L.-O., *Eur. J. Biochem.* **93**, 1 (1979).

Höök, M., Björk, I., Hopwood, J. J. and Lindahl, U., *FEBS Lett.* **66**, 90 (1976).

Hopwood, J. J., Höök, M., Linker, A. and Lindahl, U., *FEBS Lett.* **69**, 51 (1976).

Horner, A. A., *J. Biol. Chem.* **246**, 231 (1971).

Horner, A. A., *Proc. Natl. Acad. Sci. USA* **69**, 3469 (1972).

Horner, A. A., *FEBS Lett.* **46**, 166 (1974).

Horner, A. A. and Young, E., in *Glycoconjugates*, R. Schauer, P. Boer, E. Buddecke, M. F. Kramer, J. F. G. Vliegenthart and H. Wiegandt (eds.), Georg Thieme Publishers, Stuttgart, p63 (1979).

Howell, W. H., *Am. J. Physiol.* **71**, 553 (1925).

Iverius, P.-H., *Biochem. J.* **124**, 677 (1971).

Iverius, P.-H., *J. Biol. Chem.* **247**, 2607 (1972).

Iverius, P.-H., *Ciba Found. Symp.* **185**, 185 (1973).

Iverius, P.-H., Lindahl, U., Egelrud, T. and Olivecrona, T., *J. Biol. Chem.* **247**, 6610 (1972).

Jacobsson, K.-G. and Lindahl, U., *Thromb. Haemostas.* **42**, 84 (1979).

Jaques, L. B. and Waters, E. T., *J. Physiol.* **99**, 454 (1941).

Johnson, E. A. and Mulloy, B., *Carbohydr. Res.* **51**, 119 (1976).

Jordan, R., Beeler, D. and Rosenberg, R. D., *J. Biol. Chem.* **254**, 2902 (1979).

Jörnwall, H., Fish, W. W. and Björk, I., *FEBS Lett.* **106**, 358 (1979).

Kjellén, L., Oldberg, Å., Rubin, K. and Höök, M., *Biochem. Biophys. Res. Commun.* **74**, 126 (1977).

Korn, E. D., *J. Biol. Chem.* **215**, 1 (1955).

Kosakai, M., Yamauchi, F. and Yosizawa, Z., *J. Biochem.* **83**, 1567 (1978).
Kowalski, S. and Finlay, T. H., *Thromb. Res.* **14**, 387 (1979).
Krauss, R. M., Windmueller, H. G., Levy, R. I. and Fredickson, D. S., *J. Lipid. Res.* **14**, 286 (1973).
Kudryashov, B. A., Lyapina, L. A. and Ul'yanov, A. M., *Biochemistry* **42**, 346 (1977).
Kurachi, K., Schmer, G., Hermodson, M. A., Teller, D. C. and Davie, E. W., *Biochemistry* **15**, 368 (1976).
Lages, B. and Stivala, S. S., *Biopolymers* **12**, 961 (1973).
Lahiri, B., Rosenberg, R. D., Bagdasarian, A., Mitchell, B., Talmo, R. C. and Colman, R. W., *Fed. Proc.* **33**, 642 (1974).
Lam, L. H., Silbert, J. E. and Rosenberg, R. D., *Biochem. Biophys. Res. Comm.* **69**, 570 (1976).
Lane, J. L., Bird, P. and Rizza, C. R., *Br. J. Haematol.* **30**, 103 (1975).
Lane, D. A., MacGregor, I. R., Michalski, R. and Kakkar, V. V., *Thromb. Res.* **12**, 257 (1978).
Lasker, S. E. and Stivala, S. S., *Arch. Biochem. Biophys.* **115**, 360 (1966).
Lasker, S. E. and Stivala, S. A., *Nature* **213**, 804 (1967).
Laskowski, M. Jr., and Sealock, R. W., in *The Enzymes*, P. D. Boyer (ed.), Academic Press, New York, Vol. **3**, 375 (1971).
Laurent, T. C., *Arch. Biochem. Biophys.* **92**, 224 (1961).
Laurent, T. C., Tengblad, A., Thunberg, L., Höök, M. and Lindahl, U., *Biochem. J.* **175**, 691 (1978).
Li, E. H., Fenton, J. W. and Feinman, R. D., *Arch. Biochem. Biophys.* **175**, 153 (1976).
Li, E. H., Orton, C. and Feinman, R. D., *Biochemistry* **13**, 5012 (1974).
Liberti, P. and Stivala, S. S., *Arch. Biochem. Biophys.* **119**, 510 (1967).
Lindahl, U., Bäckström, G., Höök, M., Thunberg, L., Fransson, L.-Å. and Linker, A., *Proc. Natl. Acad. Sci. USA* **76**, 3198 (1979).
Lindon, J., Rosenberg, R., Merrill, E. and Salzman, E., *J. Lab. Clin. Med.* **91**, 47 (1978).
Lonky, S. A., Marsh, J. and Wohl, H., *Biochem. Biophys. Res. Comm.* **85**, 1113 (1978).
Loos, M., Volanakis, J. E. and Stroud, R. M., *Immunochem.* **13**, 257 (1976a).
Loos, M., Volanakis, J. E. and Stroud, R. M., *Immunochem.* **13**, 789 (1976b).
MacGregor, I. R., Lane, D. A. and Kakkar, V. V., *Bkochem. Soc. Trans.* **6**, 214 (1978).
Machovich, R., *Biochim. Biophys. Acta* **412**, 13 (1975).
Machovich, R., *Thromb. Res.* **9**, 123 (1976).
Machovich, R. and Arányi, P., *Biochem. J.* **173**, 869 (1978).
Machovich, R., Blaskó, G. and Arányi, P., *Thromb. Res.* **7**, 25 (1975a).
Machovich, R., Blaskó, G. and Borsodi, A., *Thromb. Haemostas.* **36**, 503 (1976).
Machovich, R., Blaskó, G., Himer, A. and Szikla, K., *Thromb. Res.* **7**, 305 (1975b).
Machovich, R., Blaskó, G. and Pálos, L. A., *Biochim. Biophys. Acta* **379**, 193 (1975c).
Machovich, R., Staub, M. and Patthy, L., *Eur. J. Biochem.* **83**, 473 (1978).
Mahley, R. W. and Innerarity, T. L., *J. Biol. Chem.* **252**, 3980 (1977).
Marciniak, E., *J. Lab. Clin. Med.* **84**, 344 (1974).
Marciniak, E., *Thromb. Diath. Haemorrh.* **34**, 748 (1975).
Marciniak, E., *Thromb. Haemostas.* **38**, 486 (1977).
Monkhouse, F. C., Frances, E. S. and Seegers, W. H., *Circ. Res.* **3**, 397 (1955).

Mookerjea, S., *Can. J. Biochem.* **56**, 746 (1978).

Moroi, M. and Aoki, N., *J. Biol. Chem.* **251**, 5956 (1976).

Nagasawa, K., Tokuyasu, T. and Inoue, Y., *J. Biochem.* **81**, 989 (1977).

Nakashima, Y., Di Ferrante, N., Jackson, R. L. and Pownall, H. J., *J. Biol. Chem.* **250**, 5386 (1975).

Niewiarowski, S., Lipinski, B., Farbiszewski, R. and Poplawski, A., *Experienta* **24**, 343 (1968).

Niewiarowski, S., Rucinski, B., James, P. and Lindahl, U., *FEBS Lett.* **102**, 75 (1979).

Nilsson-Ehle, P., Garfinkel, A. S. and Schotz, M. C., *Biochim. Biophys. Acta* **431**, 147 (1976).

Nishida, T. and Cogan, U., *J. Biol. Chem.* **245**, 4689 (1970).

Nordenman, B. and Björk, I., *Thromb. Res.* **11**, 799 (1977).

Nordenman, B. and Björk, I., *Biochemistry* **17**, 3339 (1978a).

Nordenman, B. and Björk, I., *Thromb. Res.* **12**, 755 (1978b).

Nordenman, B. and Björk, I., (in press) (1979).

Nordenman, B., Nyström, C. and Björk, I., *Eur. J. Biochem.* **78**, 195 (1977).

Nordenman, B., Danielsson, Å. and Björk, I., *Eur. J. Biochem.* **90**, 1 (1978).

Odergård, O. R. and Lie, M., *Thromb. Res.* **12**, 697 (1978).

Ogston, D., Murray, J. and Crawford, G. P., *Thromb. Res.* **9**, 217 (1976).

Olivecrona, T. and Bengtsson, G., in *Int. Conference on Athleosclerosis*, L. A. Carlson, R. Pooletti and G. Weber (eds.), Raven Press, New York, p.153 (1978).

Olivecrona, T., Bengtsson, G., Marklund, S. E., Lindahl, U. and Höök, M., *Fed. Proc.* **36**, 60 (1977).

Orakzai, S. A. and Machovich, R., *Thromb. Res.* **10**, 813 (1977).

Osler, A. G., Randall, H. G., Hill, B. M. and Ovary, Z., *J. Exp. Med.* **110**, 311 (1959).

Owen, W. G., *Biochim. Biophys. Acta* **405**, 380 (1975).

Owen, W. G., *Biochim. Biophys. Acta* **494**, 182 (1977).

Owen, W. G., Esmon, C. T. and Jackson, C. M., *J. Biol. Chem.* **249**, 594 (1974).

Pan, Y. T., Kruski, A. W. and Elbein, A. D., *Arch. Biochem. Biophys.* **189**, 231 (1978).

Pepper, D. S., Banhegyi, D. and Cash, J. D., *Thromb. Haemostas.* **38**, 494 (1977).

Petersen, T. E., Dudek-Wojciechowska, G., Sottrup-Jensen, L. and Magnusson, S., in *The Physiological Inhibitors of Blood Coagulation and Fibrinolysis*, D. Collen, B. Wiman and M. Verstraete (eds.), Elsevier/North Holland Biomedical Press, Amsterdam, p43 (1979).

Piepkorn, M. W., Schmer, G. and Lagunoff, D., *Thromb. Res.* **13**, 1077 (1978).

Pomerantz, M. W. and Owen, W. G., *Biochim. Biophys. Acta* **535**, 66 (1978).

Raepple, E., Hill, H. U. and Loos, M., *Immunochem.* **13**, 251 (1976).

Rent, R., Myhrman, R., Fiedel, B. A. and Gewurz, H., *Clin. Exp. Immunol.* **23**, 264 (1976).

Riesenfeld, J., Höök, M., Björk, I., Lindahl, U. and Ajaxon, B., *Fed. Proc.* **36**, 39 (1977).

Riesenfeld, J., Höök, M. and Lindahl, U. (in press).

Riley, J. F., *Ann. N.Y. Acad. Sci.* **103**, 151 (1963).

Rocha e Silva, M., in *Current Concepts in Kinin Research*, G. L. Haberland and U.

Hamberg (eds.), Pergamon Press, Oxford, New York, Toronto, Sydney, Paris, Frankfurt, p7 (1978).

Rodriguez, H. J. and Vanderwielen, A. J., *J. Pharm. Sci.* **68**, 588 (1979).

Rosenberg, R. D., *Fed. Proc.* **36**, 10 (1977).

Rosenberg, R. D., in *The Chemistry and Physiology of the Human Plasma Proteins*, D. H. Bing (ed.), Pergamon Press, New York, Oxford, Toronto, Sydney, Paris, Frankfurt, p353 (1978).

Rosenberg, R. D., Armand, G. and Lam, L., *Proc. Natl. Acad. Sci. USA* **75**, 3065 (1978).

Rosenberg, R. D. and Damus, P. S., *J. Biol. Chem.* **248**, 6490 (1973).

Rosenberg, R. D., Jordan, R. E., Favreau, L. V. and Lam, L. H., *Biochem. Biophys. Res. Commun.* **86**, 1319 (1979).

Rosenberg, R. D., Jordan, R., Oosta, G. M. and Gardner, W., in *The Regulation of Coagulation*, Mann/Taylor (eds.), Elsevier North Holland, Amsterdam p607 (1980).

Rosenberg, R. D. and Lam, L. H. *Proc. Natl. Acad. Sci. USA* **76**, 1218 (1979).

Rosenberg, J. S., McKenna, P. W. and Rosenberg, R. D., *J. Biol. Chem.* **250**, 8883 (1975).

Ross, R., *Ann. Rev. Med.* **30**, 1 (1979).

Ross, R. and Harker, L., *Science* **193**, 1094 (1976).

Salzman, E. W., Rosenberg, R. D., Smith, M. H., Lindon, J. N. and Favreau, L., *J. Clin, Invest.* **65**, 64 (1980).

dos Santos, E., Rothschild, H. A. and Rocha e Silva, M., in *Current Concepts in Kinin Research*, G. L. Haberland and U. Hamberg (eds.), Pergamon Press, Oxford, New York, Toronto, Sydney, Paris, Frankfurt p145 (1978).

Scanu, A. M., in *The Biochemistry of Atherosclerosis*, A. M. Scanu, R. W. Wissler, G. S. Getz (eds.), Marcel Dekker, New York and Basel (1977).

Seegers, W. H. and Marciniak, E., *Nature* **193**, 1188 (1962).

Semeraro, N., Colucci, M., Telesforo, P. and Collen, D., *Brit. J. Haematol.* **39**, 91 (1978).

Shanberge, J. N., Kamkayashi, J. and Nakagawa, M., *Thromb. Res.* **9**, 595 (1976).

Shelburne, F. A. and Quarfordt, S. H., *J. Clin. Invest.* **60**, 944 (1977).

Smith, G. F., *Biochem. Biophys. Res. Commun.* **77**, 111 (1977).

Srinivasan, S. R., Dolan, P., Radhakrishnamurthy, B. and Berenson, G. S., *Atherosclerosis* **16**, 95 (1972).

Srinivasan, S. R., Radhakrishnamurthy, B. and Berenson, G. S., *Arch. Biochem. Biophys.* **170**, 334 (1975).

Srinivasan, S. R., Radhakrishnamurthy, B. and Berenson, G. S., in *Atherosclerosis. Metabolic, Morphologic, and Clinical Aspects*, G. W. Manning and M. D. Haust (eds.), Plenum Press, New York and London (Adv. Exp. Med. Biol. Vol. 82) p155 (1978).

Stathakis, N. E. and Mosesson, M. W., *J. Clin. Invest.* **60**, 855 (1977).

Stead, N., Kaplan, A. P. and Rosenberg, R. D., *J. Biol. Chem.* **251**, 648 (1976).

Stivala, S. S., Yuan, L., Ehrlich, J. and Liberti, P. A., *Arch. Biochem. Biophys.* **122**, 32 (1967).

Stocks, J. and Galton, D. J., *Biochem. Soc. Trans.* **6**, 102 (1978).
Strunk, R. and Colten, H. R., *Clin. Immunol. Immunopathol.* **6**, 248 (1976).
Stürzebecher, J., *Act. Biol. Med. Germ.* **36**, 1893 (1978).
Stürzebecher, J. and Markwardt, F., *Thromb. Res.* **11**, 835 (1977).
Takeda, Y. and Kobayashi, N., *Thromb. Haemostas.* **38**, 684 (1977).
Taylor, R. L., Shively, J. E., Conrad, H. E. and Cifonelli, J. A., *Biochemistry* **12**, 3633 (1973).
Teien, A. N., Abilgaard, U. and Höök, M., *Thromb. Res.* **8**, 859 (1976).
Thunberg, L., Lindahl, U., Tengblad, A., Laurent, T. C. and Jackson, C. M., *Biochem. J.* **181**, 241 (1979).
Trowbridge, A. A., Caraveo, J., Green, J. B., Amaral, B. and Stone, M. J., *Am. J. Med.* **65**, 277 (1978).
Twu, J.-S., Garfinkel, A. S. and Schotz, M. C., *Atherosclerosis* **24**, 119 (1976).
Venneröd, A. M. and Laake, K., *Thromb. Res.* **7**, 223 (1975).
Villaneuva, G. B. and Danishefsky, I., *Biochem. Biophys. Res. Commun.* **74**, 803 (1977).
Villaneuva, G. B. and Danishefsky, I., *Biochemistry* **18**, 810 (1979).
Waite, M. and Sisson, P., *J. Biol. Chem.* **248**, 7201 (1973).
Walker, F. J. and Esmon, C. T., *Thromb. Res.* **14**, 219 (1979a).
Walker, F. J. and Esmon, C. T., *Biochim. Biophys. Acta* **585**, 405 (1979b).
Waugh, D. F. and Fitzgerald, M. A., *Am. J. Physiol.* **184**, 627 (1956).
Weiler, J. M., Yurt, R. W., Fearon, D. T. and Austen, K. F., *J. Exp. Med.* **147**, 409 (1978).
Wessler, S. and Yin, E. T., *Thromb. Diath. Haemorrh.* **32**, 71 (1974).
Yin, E. T., *Thromb. Diath. Haemorrh.* **33**, 43 (1974).
Yin, E. T. and Wessler, S., *Biochim. Biophys. Acta* **201**, 387 (1970).
Yin, E. T., Wessler, S. and Stoll, P. J., *J. Biol. Chem.* **246**, 3694 (1971).
Yosizawa, Z., Kotoku, T., Yamauchi, P. and Matsuno, M., *Biochim. Biophys. Acta* **141**, 358 (1967).
Yue, R. H., Starr, T. and Gertler, M. M., *Thromb. Haemostas.* **38**, 202 (1977).
Yurt, R. W., Leid, R. W. Jr., and Austen, K. F., *J. Biol. Chem.* **252**, 518 (1977).
Zucker, M. B., *Fed. Proc.* **36**, 47 (1977).

Note in Added Proof: It has recently been suggested that 3–O–sulphated glucosamine may be a unique component of the antithrombin-binding fragment of high-affinity heparin (Lindahl, U., Bäckström, G., Thunberg, L. and Leder, I. G., manuscript in preparation). This has been shown on the basis of (1) incubation of a partial-deaminative cleavage product from high-affinity heparin with 3–O–sulphatase from human urine which resulted in desulphation, and (2) isolation of a 3–O–sulphated 2,5-anhydromannitol derivative by degradation of high-affinity heparin with nitrous acid with subsequent reduction, tetrasaccharide isolation by gel chromatography and release of labelled end-group by periodate-alkali treatment. Similar treatment of an analogous tetrasaccharide derived from low-affinity heparin failed to produce the 3–O–sulphated derivative.

CHAPTER 8

Concluding remarks

Heparin remains a 'structure in search of a function'. The dichotomous approach to heparin action has yielded the realization of the generally active, occasionally specific action of exogenous heparin as compared to the relatively obscure activity of endogenous heparin (and heparan sulphate).

It is clear that heparin and related polysaccharides may partake in a variety of reactions. It is also evident that they may take on an apparent multitude of primary structures. This structural heterogeneity does not appear critical for the integrity of the tertiary structural level of the whole molecule, but rather gives rise to a multi-functional molecule in the context of *in vitro* interactions. From a clinical viewpoint, it is important that the potential exists for tailoring these molecules into highly efficient monofunctional species (for example, the separation of antilipemic and anticoagulant activities).

In more general terms, how do we proceed in translating the disaccharide code in terms of information that may have reached its 'full' meaning in a biological system as compared to a more dynamic system where many choices for generating new information may lie dormant within the molecule? The focus of these questions lies in what is presently a philosophical problem of understanding how information evolves in biological macromolecules. This is in contrast to current investigations of heparin-like polysaccharides where attempts have only been directed at establishing what information is present in these molecules so that assignments of structure–function can be immediately placed. To be sure, this approach assumes that an active structure–function principle exists for these molecules!

In this context, it is worthwhile to pursue the ramifications of a structure–function concept as applied to these molecules. It is anticipated that difficulties and complexities arise when we attempt to stratify the biological system into meaningful levels of aggregation and to generate information about how 'structural parts' are related to the 'functional whole' at each step of the hierarchical structure and the co-operative phenomena associated with them. While we necessarily make operational definitions in describing a system, such as equating structural information to the various chemical moieties on the sugar residues in heparin-like polysaccharides, to what level does this information have importance? Or more pertinently, at what functional level in a biological system can we legitimately relate such structural information? The intriguing complexity of biological systems is made clear by Weiss (1961) when he states that there is 'cellular control of molecular activities' as well as 'molecular control of cellular activities' and furthermore 'organismal control of cellular activities' and so on, i.e. function, albeit equated with relatively high orders of hierarchy, may determine structure and vice-versa. Eigen (1971) views the hierarchical system as a series of closed loops being evolution-generated by random events which do not have any structural significance but he points to the importance of how certain random events are able to 'feed back to their origin and thus become the cause of some amplified action ... which results in the build up of macroscopic functional organizations which include self-reproduction, selection and evolution to a level of sophistication where the system can escape the prerequisites of its origin and change the environment to its own advantage'.

In terms of a functional level in biological systems, the ability of a system to reproduce and adapt is paramount. One can speak of the evolution of molecules, macromolecules, organelles, cells or organs only as they relate to the reliability of adaptation and reproduction for the organism as a whole. What is required is how do macromolecules such as heparin, assure their own reproduction and reproduction of the organism as a whole. What closed loop systems are these molecules associated with or are they searching for other arenas of action?

A general view emerges that hierarchical structures and aggregates produced by evolutionary and adaptive processes are the result of the interplay of two processes

1) genetically determined specialization that essentially reflects the historical circumstance of the system

2) direct physiological adjustment or environmental plasticity or inductive development.

The combination of these two factors exists at every level of organization although in different proportions. The effect of these evolutionary and adaptive components on glycosaminoglycan and proteoglycans is of particular interest. Mathews (1975), in an excellent treatise on the evolutionary aspects of connective tissues, has emphasized that direct physiological adjustment is clearly illustrated in connective tissues. The response to environmental influences appears to be an essential property of supporting tissues and that this characteristic is derived largely from the capacity for fine modulation of the structure of its constituent macromolecules. Mathews continues 'it is revealed for example, in the *regulation* of supramolecular structures via variation of a large number of secondary chemical features of secreted macromolecules and, generally, in the responsiveness and functional adaptation of cells and tissues to chemical and physical environmental factors'. In specific terms, the proteoglycans while consisting of polypeptides and polysaccharide moieties represents a heterogeneous situation in terms of their relation to the gene; while the polypeptide moieties are directly coded, the polysaccharide moieties are indirectly coded through the regulatory enzymes synthesizing them. It is principally through this secondary chemical modification of primary directed coded structures that a very great number of different molecular forms is achieved by each organism. These modifications include

1) the size and number of glycosaminoglycans chains attached to the polypeptide moiety;

2) extent and site of sulphation;

3) degree of epimerization of the glucuronic acid residue to form iduronic acid. The fact that the glycosaminoglycans and their embellishments are then distant from the genes suggests that they are primary sensors for adaptive processes involved and thereby control biosynthesis and functional inter-relationships rather than

the genes themselves — a general principle outlined by Simpson (1964).

From an evolutionary point of view the glycosaminoglycans represent a special case in point. They represent one of the most primitive groups of high polymers, although Dietrich and co-workers (1977) have found that only hyaluronate seems to be present in bacteria and protozoa and that sulphated glycosaminoglycans seems to be absent. On this basis, several investigators (Chiarugi, 1976; Dietrich *et al.*, 1977) have suggested the direct relationship between the emergence of the sulphated glycosaminoglycans and the emergence of tissue-organized life forms where cell–cell interaction and production of intercellular matrices are involved.

REFERENCES

Chiarugi, V. P., *Exp. Cell. Biol.* **44**, 251 (1976).

Dietrich, C. P., Sampaio, L. O., Toledo, O. M. S. and Cássaro, C. M., *Biochem. Biophys. Res. Comm.* **75**, 329 (1977).

Eigen, M., *Quart. Rev. Biophys.* **4**, 149 (1971).

Mathews, M. B., *Connective Tissue. Macromolecular Structure and Evolution*, Molecular Biology, Biochemistry and Biophysics, Vol. 19, Springer-Verlag, Berlin, Heidelberg and New York (1975).

Weiss, P. A., in *Molecular Control of Cellular Activity*, J. M. Allen (ed.), McGraw-Hill, New York, p1 (1961).

Appendix 1

MOLECULAR WEIGHTS AND STRUCTURES OF MONOSACCHARIDE UNITS

Structure	*Description*	*Molecular Weight*
CH_2OH, O, OH, OH OH, $HNCOCH_3$	α- D-N-acetyl glucosamine	221
CH_2OH, O, OH, OH OH, $HNSO_3^-$	α-D-N-sulphated glucosamine	258
$CH_2OSO_3^-$, O, OH, OH OH, $HNCOCH_3$	α- D-6-sulphated N-acetyl glucosamine	300
CH_2OH, O, OH, OH OH, $HNSO_3^-$	α-D-di N-6-sulphated glucosamine	347
CO_2^-, O, OH, OH OH, OH	α- L-iduronate	193
CO_2^-, O, OH, OH OH, OSO_3^-	α- L-2-sulphated iduronate	272

Structure	Description	Molecular Weight
CH_2OH, OH, OH, CHO	2,5-anhydro-D-mannose	162
$CH_2OSO_3^-$, OH, OH, CHO	2,5-anhydro-D-mannose 6-sulphate	241

Appendix 2

PUBLISHED VALUES OF PARTIAL SPECIFIC VOLUMES AT 25°C OF COMMERCIAL HEPARIN PREPARATIONS

Description	$\bar{v}$ *(ml/g)*	*Solvent*	*Reference*
NI	0.479	1MKCl	Patat & Elias (1959)
Riker (mucuos)	0.46	1MNaCl	Braswell (1968)
Abbot (pork slime)	0.41	1MNaCl	Braswell (1968)
Organon 13127 (lung	0.38	1MNaCl	Braswell (1968)
Organon 9986 (lung)	0.42	1MNaCl	Braswell (1968)
Boots (ox lung)	0.42	1MNaCl	Braswell (1968)
Lung Heparin by-products 0.5	0.56	0.2M NaCl	Linker & Hovingh (1973)
Lung Heparin by-products 0.9	0.53	0.2M NaCl	Linker & Hovingh (1973)
Lung Heparin by-products 1.2	0.45	0.2M NaCl	Linker & Hovingh (1973)
Lung Heparin by-products 1.4	0.50	0.2M NaCl	Linker & Hovingh (1973)
Organon 15595 (lung)	0.47	0.5M NaCl	Lasker & Stivala (1966)
Organon 15595 (lung)	0.47	1.0M NaCl	Lasker & Stivala (1966)
AB Vitrium	0.42	0.2M NaCl	Laurent (1961)
Organon 22419 (lung)	0.44–0.49	Varying [Ca^{2+}] in 0.1M Tris buffer, pH 7.5	Lages & Stivala (1973)

References

Braswell, E., *Biochim. Biophys. Acta* **158**, 103 (1968).
Lages, B. and Stivala, S. S., *Biopolymers* **12**, 127 (1973).
Lasker, S. E. and Stivala, S. S., *Arch. Biochem. Biophys.* **115**, 360 (1966).
Laurent, T. C., *Arch. Biochem. Biophys.* **92**, 224 (1961).
Linker, A. and Hovingh, P., *Carbohydr. Res.* **29**, 41 (1973).
Patat, F. and Elias, H. G., *Naturwissenschafen* **46**, 322 (1959).

Abbreviations

NI = not identified

Appendix 3

PUBLISHED VALUES OF SPECIFIC REFRACTIVE INDEX INCREMENTS (*dn/dc*) OF COMMERCIAL HEPARIN PREPARATIONS

Description	*dn/dc (ml/g)*	*Wavelength (nm)*	*Solvent*	*Reference*
NI	0.121	546	1MKCl	Patat & Elias (1959)
Riker (mucuos)	0.136	NI	1MNaCl	Braswell (1968)
Abbot (pork slime)	0.135	NI	1MNaCl	Braswell (1968)
Organon 13127 (lung)	0.139	NI	1MNaCl	Braswell (1968)
Organon 9986 (lung)	0.138	NI	1MNaCl	Braswell (1968)
Boots (ox lung)	0.140	NI	1MNaCl	Braswell (1968)
NI	0.128	436	H_2O	Loucas et al. (1971)
NI	0.129	436	0.05M$CaCl_2$	Loucas et al. (1971)
NI	0.123	436	0.1M$CaCl_2$	Loucas et al. (1971)
NI	0.100	436	1.16M$CaCl_2$	Loucas et al. (1971)

References

Braswell, E., *Biochim. Biophys. Acta* **158**, 103 (1968).

Loucas, S. P., Keh, K. C. and Haddad, H. M., *J. Pharm. Sci.* **60**, 1109 (1971).

Patat, F. and Elias, H. G., *Naturwissenschafen* **46**, 322 (1959).

Abbreviations

NI = not identified

Appendix 4

LIST OF ABBREVIATIONS AND SIGNS

(see also Tables 1.1 and 7.4, and Figure 3.2)

ADP	adenosine di-phosphate
AMP	adenosine mono-phosphate
APTT	activated partial thromboplastin time
AT	antithrombin III
AT_m	modified antithrombin
anh-Man$(SO)_3^-$	anhydromannose (sulphate)
arg	arginine
A	equivalent freely-jointed segment length
a_m	distance with which centre of a micro-ion can approach the centre of a cylinder
BP units	units of anticoagulant activity based on British Pharmacopoeia standard
b	contour length of polyion carrying unit charge
ChTg	Chymotrypsinogen
C_i	concentration of ion i
c	polymer concentration
D	diffusion coefficient
D_0	diffusion coefficient at infinite dilution
DEAE	Diethyl amino ethyl (cellulose)
D	dielectric constant
EDTA	ethylene diamine tetraacetate
e	electron charge
Gal	galactose
GlcN	glucosamine
GlcNAc	N-acetyl glucosamine
$GlcNSO_3^-$	N-sulphate glucosamine
$6\text{–O–}SO_3^-$ | $GlcNSO_3^-$	6–O–sulphate, N-sulphate glucosamine
GlcUA	glucuronic acid
Gly	glycine
GuHCl	guanidiium hydrochloride
HA	high-affinity (in relation to heparin binding to antithrombin III)
Hi	histidine
I_{50}	concentration of inhibitor required for 50% inhibition
IdUA	iduronic acid

$IdUASO_3^-$ (also IdUA | 2–O–SO_3^-)	2–0–sulphated iduronic acid
Ile	isoleucine
IU units	units of anticoagulant activity based on International standard
I	ionic strength
J_{ik}	^{1}H-nmr coupling constants
K_{ass}	association constant
K_{av}	partition coefficient on gel chromatography
$K_0(x)$, $K_1(x)$	modified Bessel functions of the second kind
kat (=katal)	thrombin activity corresponding to conversion of 1 mol of synthetic chromogenic substrate per sec
k	Boltzmannis constant
LA	low affinity (in relation to heparin binding to antithrombin III)
leu	leucine
lys	lysine
l	length
$[M]_D$	optical rotation at 589nm (deg/mol disaccharide unit)
M	molecular weight
M_n	number average molecular weight
M_w	weight average molecular weight
dn/dc	specific refractive increment
N	Avogadro's number
n	degree of polymerization of disaccharide units
n_p	concentration of polyion in equiv/L
PAPS	adenosine 3′-phosphate-5′-sulphato phosphate
Phe	phenylalanine
Pro	proline
$(\bar{r}^2)^{\frac{1}{2}}$	root-mean-square end-to-end distance
r_c	cylindrical radius
R_g	radius of gyration
SDS	sodium dodecyl sulphate
Ser	serine
S	sedimentation coefficient
S_0	sedimentation coefficient at infinite dilution
T	thrombin
t	time
Tris	tris (hydroxymethyl)) amino methane
T	temperature
UA	hexuronic acid
UDP	uridine di-phosphate
USP units	units of anticoagulant activity based on US Pharmacopoeia standard
$\bar{v}$	partial specific volume
W_{el}	electrostatic work
Xyl	Xylose
Z	charge valence
α	degree of dissociation

$[\alpha]^T$	optical rotation at 589nm and temperature T (deg g^{-1})
β	charge contained in a unit length of cylindrical polyion
γ_{Na^+}	sodium-ion activity coefficient
$\Delta^{4,5}$	unsaturated bond between C-4 and C-5 on glycose residue
δ_i	^{1}H-nmr chemical shifts
ε	extinction coefficient
ζ	thrombin inactivating activity
$[\eta]$	intrinsic viscosity
$[\eta]_\infty$	shielded viscosity
η_0	solvent viscosity
η_{sp}	specific viscosity
θ	molar ellipticity
κ	Debye-Hückel parameter (Eq. 3.2)
ν_{as}	infra-red antisymmetric stretching vibration
ν_s	infra-red symmetric stretching vibration
ξ	charge density parameter

Subject Index

All subjects are prefaced by Heparin (and Related Polysaccharides) unless otherwise indicated.